INSTRUCTOR'S MATHEMATICA® MANUAL

LYLE COCHRAN
Whitworth College

to accompany

Update

LINEAR ALGEBRA AND ITS APPLICATIONS

SECOND EDITION

DAVID C. LAY
University of Maryland

ADDISON-WESLEY

An imprint of Addison Wesley Longman, Inc.

Reading, Massachusetts • Menlo Park, California • New York • Harlow, England
Don Mills, Ontario • Sydney • Mexico City • Madrid • Amsterdam

Reproduced by Addison Wesley Longman from camera-ready copy supplied by the author.

ISBN: 0-201-87489-X

1 2 3 4 5 6 7 8 9 10 CRC 03 02 01 00 99

TABLE OF CONTENTS

FOREWORD

This manual is for any instructor who is using Mathematica and *Linear Algebra and Its Applications* together for the first time. It will greatly simplify your task of combining Mathematica with the text, because it is written by a colleague who has done this with considerable success. This manual carefully describes everything you need to know about planning and conducting the course.

The author, Professor Lyle Cochran has been using Mathematica for about eight years, in a variety of courses. Over the past three years, he has developed a series of Mathematica notebooks that his students have thoroughly enjoyed. He has developed palettes for the notebooks that greatly simplify the use of Mathematica in linear algebra, allowing students to focus on mathematics rather than Mathematica.

I am pleased and grateful that Professor Cochran has created such interesting projects and written this informative manual. I am sure that both you and your students will appreciate his work.

David C. Lay
University of Maryland

INTRODUCTION

I have been using Mathematica in linear algebra and several other courses for a number of years now. Before version 3.0 of Mathematica was released in 1996, many of my students would struggle with the syntax of the Mathematica language. For example, Mathematica commands are case-sensitive and begin with a capital letter. Therefore, when students would type in a command from the keyboard, they would often forget to capitalize the first letter in a command, leading to undesired output. Syntax errors such as these would leave some students frustrated and discouraged with Mathematica. *However these student syntax errors were almost completely eliminated when I starting incorporating palettes into computer assignments in the 1997.* Palettes, new to Mathematica 3.0, are similar to a group of calculator buttons, providing users with shortcuts to typing in commands from the keyboard. Furthermore, palettes are customizable so that I can create a palette containing only those commands needed by students to complete a particular computer assignment. So instead of typing in a command from the keyboard, my students can now simply click on the appropriate palette button to enter a command. This new feature has helped my students concentrate on the mathematical material in the computer assignments instead of Mathematica syntax.

My linear algebra class meets three times per week in a traditional classroom setting, and students complete the computer assignments outside of class in a computer laboratory on campus. I use an overhead computer projection system to give brief overviews of the computer exercises in class and I also incorporate Mathematica into some of my lectures. Most of the students are either math or computer science majors and they enjoy working through the applications introduced in the computer assignments. They also appreciate the power of Mathematica to simplify tedious or complicated calculations.

Please send me e-mail if you have any comments or questions regarding the computer assignments or other related issues.

Lyle Cochran
Dept. of Mathematics & Computer Science
Whitworth College
Spokane, Washington 99251
LCOCHRAN@WHITWORTH.EDU

1. GETTING STARTED

The supplemental computer assignments contained in this manual are based upon concepts presented in David Lay's textbook, *Linear Algebra and Its Applications, Updated Second Edition.* Each assignment requires students to solve linear algebra exercises using the computer algebra system Mathematica, preferably version 3.0 or higher.

PREPARE

If you have not used Mathematica with Lay's text, this manual will help you put them together. If you are not familiar with Mathematica, spend some time entering and executing commands before the first class session. A quick way to get started is to work through some of the computer projects in this manual and some of the [M] exercises in the text. You might also examine some of the projects on the CD that comes with the text. Pages 6-15 in this manual describe the Mathematica environment and the palettes that now make Mathematica so much easier to use. Page 23 lists some references that you may find helpful.

You can quickly get an idea of how Mathematica can be integrated into your course by looking at the Mathematica Appendix in the updated *Study Guide,* which is a supplement to Lay's text. There you will see how commands are introduced only as they are needed for linear algebra computations. (If you have not used the text before, be careful, because the order of ideas may be quite different from what you expect. Matrix multiplication, for instance, is not defined until Chapter 2.)

EVALUATE SOFTWARE AND HARDWARE REQUIREMENTS

If you plan on purchasing Mathematica and making it available for students to use, then consider buying an educational site license. A site license will allow you to install Mathematica on several machines at a discounted price. Students can also purchase *Mathematica for Students* at a small fraction of the cost of the standard version of Mathematica. *Mathematica for Students* is a fully functional version of Mathematica, which runs at the same speed and produces the same output as the standard version of Mathematica. But only full-time students enrolled at an accredited academic institution are eligible to purchase and use the student version. For more information about the student version and eligibility requirements, visit www.wolfram.com on the Web.

When an educational site license is purchased, a copy of *The Mathematica Book* (see [13]) will accompany the software. An electronic version of the entire Mathematica book can be accessed through the Help menu while running the Mathematica program. *Mathematica for Students* does not come with a hard copy of *The Mathematica Book,* but it does contain the electronic version. I prefer to use the electronic version over the hardbound copy since the electronic book is interactive. Actually, your students will seldom, if ever, need more documentation for this course than the Appendix in the *Study Guide* and perhaps the online Help menu.

Mathematica runs on most major platforms and operating systems. For those using Windows 95 or Windows NT, 116 MB of hard disk space and 16 MB of RAM are recommended. Macintosh users using the MacOS 7.1 operating system are recommended to have 118 MB of hard disk space and 12 MB of RAM. Mathematica software also runs on Linux, Unix, NEXTSTEP, IBM OS/2 and DEC OpenVMS 7.0 operating systems. For more information, visit the Mathematica web site.

ORDER STUDENT MATHEMATICA SOFTWARE AND STUDY GUIDES EARLY

Ask your bookstore to stock *Mathematica for Students,* in both the Windows and the Macintosh versions, and Lay's *Study Guide, Updated Second Edition.* Students should have these materials by the first day of class. Even if Mathematica is available in the computer labs at your school, you will still want the bookstore carry a few copies of *Mathematica for Students* since there will always be a few students who will want to purchase a copy for home use. Bookstore managers often order reduced quantities of a supplement such as the *Study Guide,* unless you make it a required item. Having the *Study Guide* available will save you the trouble of explaining how to use Mathematica with homework exercises.

INSTALL LAY'S MATHEMATICA TOOLBOX BEFORE CLASSES START

Some additional Mathematica commands and electronic data corresponding to over 800 exercises (including all of the [M] exercises) in the textbook are contained in files on the CD-ROM that comes with the textbook and on the Web site at www.laylinalgebra.com. These commands and data will be needed for homework and for some of the computer assignments.

To obtain these files, create a new folder called LayData (be sure to spell this exactly as it appears here with a capital L and a capital D) inside the Mathematica folder called ExtraPackages. For Mathematica 3.0, the path to this folder is

Wolfram Research / Mathematica / 3.0 / AddOns / ExtraPackages / LayData

To obtain the files for the LayData folder, begin by opening the Web site on the CD-ROM or on the Internet. Choose Mathematica under Download Data and press Go. Select Mthma.zip if you are using a PC or select Mthma.hqx if you are using a Macintosh. You will then see a window asking if you would like to save the file (containing the files for the LayData folder) to your disk or if you would like to open the file. Select Open and a list of files will appear. Extract the following files and place them in the LayData folder:

> LayFunctions.m, Chapter1.m, Chapter2.m, Chapter3.m, Chapter4.m, Chapter5.m, Chapter6.m, Chapter7.m, Misc.m.

The contents of the LayData folder will often be referred to as Lay's Mathematica Toolbox or as the Toolbox data.

If your department or campus has a local area network with Mathematica installed, ask the network administrator to place the Toolbox data into a new LayData folder as described above.

Once the Lay Toolbox has been installed, verify that it is operating correctly. To do this, read pages MM-1 through MM-6 in Lay's *Study Guide, Updated Second Edition*. After you start Mathematica, type in and execute the commands on lines In[1] through In[11] found on pages MM-5 and MM-6 in the *Study Guide* and make sure that the commands work correctly. If your students will be using computers on campus, go to the place where students will be doing their work and repeat the process of checking the commands in lines In[1] to In[11].

Students who purchase their own copy of Mathematica will also need to install the Toolbox on their own computers, in the ExtraPackages folder using the steps described above. Instructions for creating the LayData folder are also on the Web/CD.

INSTALL THE ELECTRONIC FILES FOR THE COMPUTER ASSIGNMENTS

Create a new folder and assign a name to the folder. Next, to obtain the computer assignments associated with this manual, log on to the Web site www.laylinalgebra.com (the files are not available on the CD-ROM). The computer assignment files are in the same file from which you extracted the Toolbox data. Extract the following files using the same process that you used the obtain the Toolbox data and place the files in the new folder:

> Intro.nb, Poly.nb, Rent.nb, Elmat.nb, Code.nb, LUfact.nb, Graphics.nb, Motion.nb, Det.nb, Markov.nb, Owls.nb.

These are the interactive computer assignments described in this manual. Hard copies of these files are located in the last section of this manual.

OBTAIN DATA FOR CASE STUDIES AND APPLICATION PROJECTS

Data files for the Case Studies and the Application Projects are contained in text files on the Web and on the CD. If you decide to assign one or more of these projects that has an accompanying data, then you will need to open the associated text file. To do this, go to the chapter in which the Case Study or Application project is located using the Application by Chapter button at the Web site. Select the Case Study or Applications button to see a list of the studies or applications contained in this chapter. When you select View PDF Now under the Case Study or Application of your choice, there will be an option Downloading Data for this Application. When you select this option, a text file will appear containing the data for that particular project. To use this data with Mathematica, you will have to copy and paste this data into a new Mathematica notebook.

CONSIDER EQUIPMENT FOR CLASSROOM DEMONSTRATIONS

Determine what computer equipment is available at your institution for classroom demonstrations. For classroom demonstrations, my department has a portable cart containing a computer, monitor and color LCD projector. The projector allows me to display the monitor images directly onto a white wall. On days when I plan on using the cart for classroom demonstrations, I arrive in class a few minutes early to set up the equipment.

If you do not have access to a projector, then, early on in the course, you may want to spend some of your scheduled office hours in the student computer lab to assist students with Mathematica. If you have a teaching assistant, then you may want to spend some time before the course begins training this person to use Mathematica and then let this person assist students in the computer lab.

2. PLANNING THE COURSE

STUDENT BACKGROUND

Before your course begins, it would be helpful to know a little bit of academic information about your students (e.g., student majors and minors, year in school,...) in order to tailor the course to fit the needs and interests of your students. Examine the computer projects contained in this manual and pick projects that you think are appropriate for the students in your course. For example, many of my linear algebra students are computer science majors and therefore I like to include the computer graphics project in my list of assigned computer exercises.

PLANNING YOUR SYLLABUS

In the *Instructor's Edition* to *Linear Algebra and Its Applications*, there is a section entitled *NOTE TO THE INSTRUCTOR* containing helpful information for course preparation. Study this section carefully and pay careful attention to the sample syllabi when designing your course. Sample syllabi, with review sheets and examinations, are available at the Web site www.laylinalgebra.com. You may also want to read [3], which describes five recommendations for a first course in Linear Algebra.

Most students will want to know how examinations, homework, class participation and computer assignments are weighted in the course. If this is your first time using Mathematica computer assignments in a course, then I would suggest making the computer assignments worth a relatively small percentage of the total possible points that can be earned in the course. For example, if homework accounts for 10% of the total possible points in the course, then let computer assignments account for 5% or 10% of the final grade. Another possibility is to include computer exercises as part of the homework points earned by the students.

STUDENT COMPUTER LAB INSTRUCTIONS

Prepare a sheet of computer lab instructions which includes the following information and pass out this information to students on the first day of class.

- Location of campus computer lab facilities.
- Instructions about obtaining and using computer log-on names and passwords.
- Hours and days of the week when computer labs are available for student use.
- Instructions on how to start up the Mathematica program.
- Instructions stating the location of the computer assignments.

USING WEB/CD RESOURCES

The CD-ROM that accompanies each student and instructor copy of *the Updated Second Edition of Linear Algebra and Its Applications* contains a Case Study for each chapter and 21 Application Projects tied to specific sections of the text. Each textbook also contains a password that gives the student or instructor access to the Web site www.laylinalgebra.com. In this manual, the term Web/CD refers to both locations for the data. Section 6 in this manual provides a brief description of all the Case Studies and Application Projects. Instructions for downloading and using the data for these projects are on the Web/CD.

For students, the web site contains a mirror copy of the CD materials, with updates and additions released after the text was published. For example, the Mathematica notebooks for the projects in this Instructor's Manual are on the web site, and can be accessed by choosing Download Data followed by Mathematica and then by following the instructions for downloading the files. Faculty have additional resources on their version of the web site, including transparency master, sample review exams, and other materials.

REASONS FOR INCLUDING COMPUTER ASSIGNMENTS

This section discusses some of the advantages of including computer exercises in an introductory linear algebra course. You are encouraged to present your rationale for assigning computer exercises in class so students do not develop the misconception that computer assignments are just "busy-work".

Computer exercises allow students to explore applications.

Most students enrolled in linear algebra are better able to grasp and retain the material if they can understand the application of the course content to "real-world" problems. The computer assignments contained in this manual include applications to curve fitting, cryptography, computer graphics, animation and population analysis.

Computer exercises reinforce concepts presented in the textbook.

Topics from Lay's text found in the computer exercises include row reduction, LU factorization, matrix multiplication, computer graphics and Markov chains.

Most applications of linear algebra in the "real world" are solved using computer software.

One of my goals for introducing computer exercises in this course is to give students exposure to using a computer as a scientific tool for solving linear algebra exercises. While many students are comfortable using the computer for accessing the Internet and for e-mail, they have had much less practice at using the computer to solve scientific problems. Completing linear algebra exercises with Mathematica will help students see the value of using a computer as a scientific tool.

The computer allows students to focus on ideas and concepts rather than tedious arithmetic computations.

For example, the row reduction algorithm can be a tedious and time-consuming process. The computer allows students to visualize the row-reduction process without having to focus on the arithmetic computations.

Most students enjoy doing computer exercises.

For many students, computer assignments are an enjoyable component of an introductory linear algebra course. For example, in one of the computer assignments, students use matrix multiplication to create computer animations and most students are fascinated to see the role of linear algebra in computer graphics.

3. ASSIGNMENTS AND PROJECTS

PREREQUISITES FOR SUCCESS

Experienced faculty have learned that the following two conditions are necessary for successful integration of computing and mathematics:

a. **Before giving a computer assignment to your students, complete all the exercises in the assignment and modify the assignment to meet the specific needs of your students.**

   It is absolutely necessary for you to work through each computer assignment, evaluating the usefulness and level of difficulty of the exercises, before assigning it to students. You are encouraged to modify the computer assignments as you see fit.

b. **Have students complete an assignment very early in the course, and then continue to require computer work (if only a few [M] exercises from the text) at regular intervals.**

   Since students will have more free time available at the beginning of the course, it is a good idea to make a computer assignment due early in the semester when they have more time to adjust to the Mathematica computing environment. Let the students know that the first computer session may not go as smoothly as they would expect since they are learning a new piece of software (although my experience has been that students usually encounter few, if any problems, when working through the first assignment). Continue to assign computer work on a regular basis so that students become familiar and comfortable working with Mathematica.

Two other strategies to maintain student interest and to improve the course as you teach it.

c. **Discuss computer assignments in class.**

   When you are ready to assign a computer project for students to complete, briefly introduce the computer assignment in class and discuss the relevance of the assignment to course topics concurrently being covered in lecture. Neglecting to discuss computer assignments may give students the misperception that computer assignments are "busy work" with no real importance to lecture and written homework assignments.

d. **Get feedback from students.**

   After students have completed an assigned computer project, I often ask them if they found the project interesting and challenging. Many, but not all, are honest about whether or not they liked the given assignment and indicate what they learned by completing the assignment. Ask students about the amount of time that it took to complete a given assignment in order to get a better sense of the workload being added to the course by the computer work.

ARRANGING THE MATHEMATICA MATERIAL

e. **Provide a brief introduction to Mathematica before the first assignment.**

   Many of the problems and difficulties that student have with Mathematica can be avoided by providing a short five-minute, in-class demonstration on running and using Mathematica. On the first or second day of classes, I use a projector to demonstrate how to open a Mathematica file and how to execute Mathematica commands. If an in-class demonstration is not feasible, some other options include (1) training a student teaching assistant to help students use Mathematica, (2) holding office hours in the computer lab early on in the semester, or (3) providing hands-on time during a regularly scheduled lecture hour to allow students to work with the software.

f. Assign [M] exercises out of Lay's textbook.

The textbook includes [M] labeled exercises. These exercises are designed to be solved with the help of a computer. I strongly encourage you to have students solve some of these exercises with Mathematica as a part of their regular homework. Electronic data corresponding to over 800 exercises in the textbook, including all of the [M] exercises, and instructions on how to use the [M] exercise data, can be downloaded off the Web/CD. The *Student Study Guide* contains an appendix of Mathematica commands and instructions, which will be a helpful reference guide when working with the [M] exercises. Assigning [M] exercises will help students become more competent and comfortable with Mathematica.

g. If this is your first time using computer assignments in a linear algebra course, then assign no more than three or four computer projects, other than [M] exercises in the text.

Focus on the *quality* of computer assignments, not quantity. It takes time and hard work on your part to prepare and customize computer assignments to meet the needs of your students.

h. Carefully evaluate the effectiveness and level of difficulty of each computer assignment before making another assignment.

If this is the first time you have assigned computer work in a linear algebra course, then you need to carefully monitor student progress on computer assignments. The following questions may prove useful in evaluating each assignment:

- *Did the computer assignment take an excessive amount of time for students to complete?*
- *Were students presented with any unexpected problems when working through an assignment in the computer laboratory?*
- *What do you need to change or modify to make the next computer assignment go more smoothly for students?*

ADMINISTRATIVE ISSUES

i. Be clear about assignment due dates.

Make sure to announce computer assignment due dates well in advance and encourage your students to start working on computer assignments early to allow extra time to deal with unexpected problems which may arise. If students complete their assignments in a school computer lab, it is best to announce the computer lab hours at the beginning of the course to avoid problems such as the lab being closed or in use for some other class. Unfortunately, some students will procrastinate and will not begin working on an assignment until the day before it is due. Inevitably, a few students will have excuses for incomplete assignments or late work, therefore you need to have a plan for dealing with late assignments in a consistent, fair manner.

j. Determine the acceptable form for completed assignments.

As will be discussed later on in this manual, the computer assignments are interactive tutorials, which if printed will require several pages of paper. If you want to save paper, there are several alternatives to having students turn in hard copies of their work. I ask my students to send in computer assignments in the form e-mail attachments. To make the assignments easier to grade electronically, I have set up each computer assignment so that all written explanations entered by students appear in the color blue. This will make it easier for you to locate student work should you decide to grade the assignments electronically. You could also type in comments using a red font or some other color so that students can easily locate your comments. Another possibility is to have students turn in a copy of their work on a floppy disk. Be sure to give clear instructions to your students about the form in which you want them to turn in their work.

k. Determine whether students will work individually or in groups.

I usually have students work on assignments individually, but many instructors prefer to require students to work in pairs or groups. Your decision may be determined by a number of factors including the number of computers available for student use, the size of your class and whether or not you have teaching assistants that can help you grade your assignments.

4. THE MATHEMATICA COMPUTING ENVIRONMENT

The purpose of this section is to introduce instructors to Mathematica *notebooks, palettes,* and *initialization cells.*

MATHEMATICA 3.0 VS. EARLIER VERSIONS

The computer assignments contained in this manual were written using Mathematica *3.0.* Earlier versions of Mathematica can be used to complete the assignments, but the following information pertaining to palettes will not be available in the earlier version. Instructors who plan on using some of the computer assignments with an earlier version of Mathematica will need to modify the electronic copies of the computer assignments before assigning them to students.

TEXT-BASED INTERFACE

When working with Mathematica, you will either use a *text-based* interface or a *notebook* interface. The computer assignments contained in this manual were created using a notebook interface and are stored in files called Mathematica *notebooks.* The notebook interface is preferred over the text-based interface and is discussed in the following section. For users limited to the text-based interface, input commands must be typed directly into the computer using only the available characters on the keyboard. A casual glance at the computer assignments will reveal that some of the text contained in the assignments could not have been created using a text-based interface.

For example, to enter a 2-by-2 matrix whose first row consists of the numbers 1 and 2 and whose second row consists of the numbers 3 and 4, text-based users must type in {{1,2},{3,4}} while users working with a notebook interface will be able to enter a matrix in a more natural way using palettes (which are described later on in this manual).

Another disadvantage of the text-based environment is that the attached computer assignments will not be interactive. Instead, users will have to type in all of the input commands from scratch whereas notebook users will be able to work with the electronic computer assignments interactively as described in the following section. Instructors then need to provide specific information and instructions to students concerning differences that will be encountered when working with the computer assignment in the text-based environment.

Instructors and students using the text-based interface will need to read the Overview of Mathematica Computer Assignments in Section 6 for special information about loading customized commands into memory at the beginning of a computer session.

NOTEBOOK INTERFACE

Instructors and students using the notebook interface will need to become familiar with Mathematica *notebooks* and *palettes.*

Mathematica Notebooks

Each computer assignment contained in this manual is a hard copy of an electronic file called a Mathematica notebook – an interactive document organized into a sequence of *cells.* Each cell contains a specific type of information such as graphics, text, input and output. Associated with each cell is a cell bracket located to the right side of the window (see Figure 1). Mathematica commands are entered into input cells and when the commands contained in the input cells are executed, the output of the commands are displayed immediately after the input cell in an output cell.

Figure 1: Example of a Mathematica Notebook

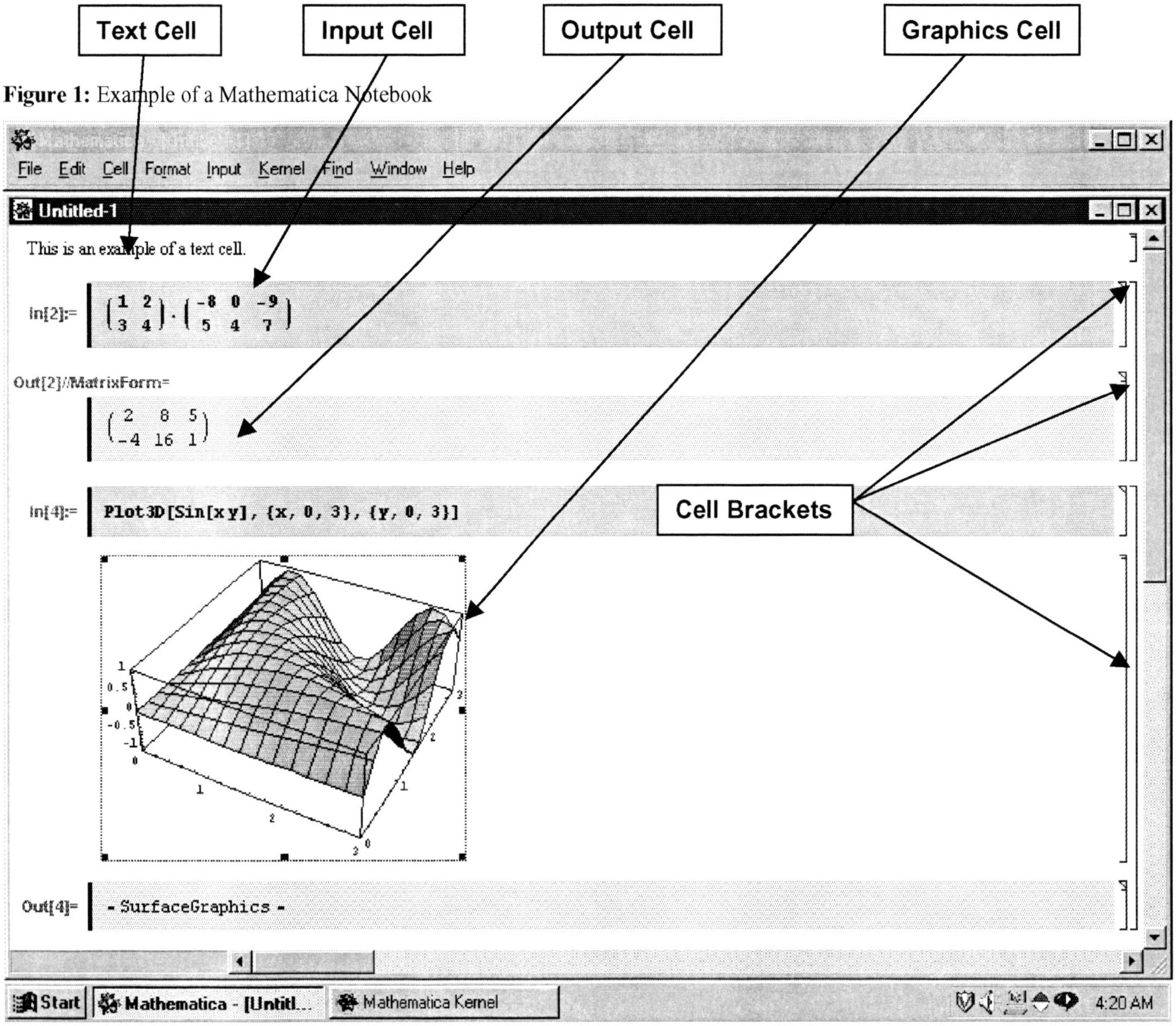

Closed Cell Groups

A group of cells can either be *closed* or *open*. When it is closed, only the first cell in the group is visible and when it is open, all the contents in all of the cells in the group are visible. An arrow (or, what to me looks like an upside-down flag) indicates a closed group of cells (see Figure 2) and the group of cells can be opened by double clicking on this arrow (see Figure 3). The electronic version of each computer assignment accompanying this manual will initially have most of its cell groups closed, so that when you first look at the notebook, you will see an outline of the contents of the computer assignment.

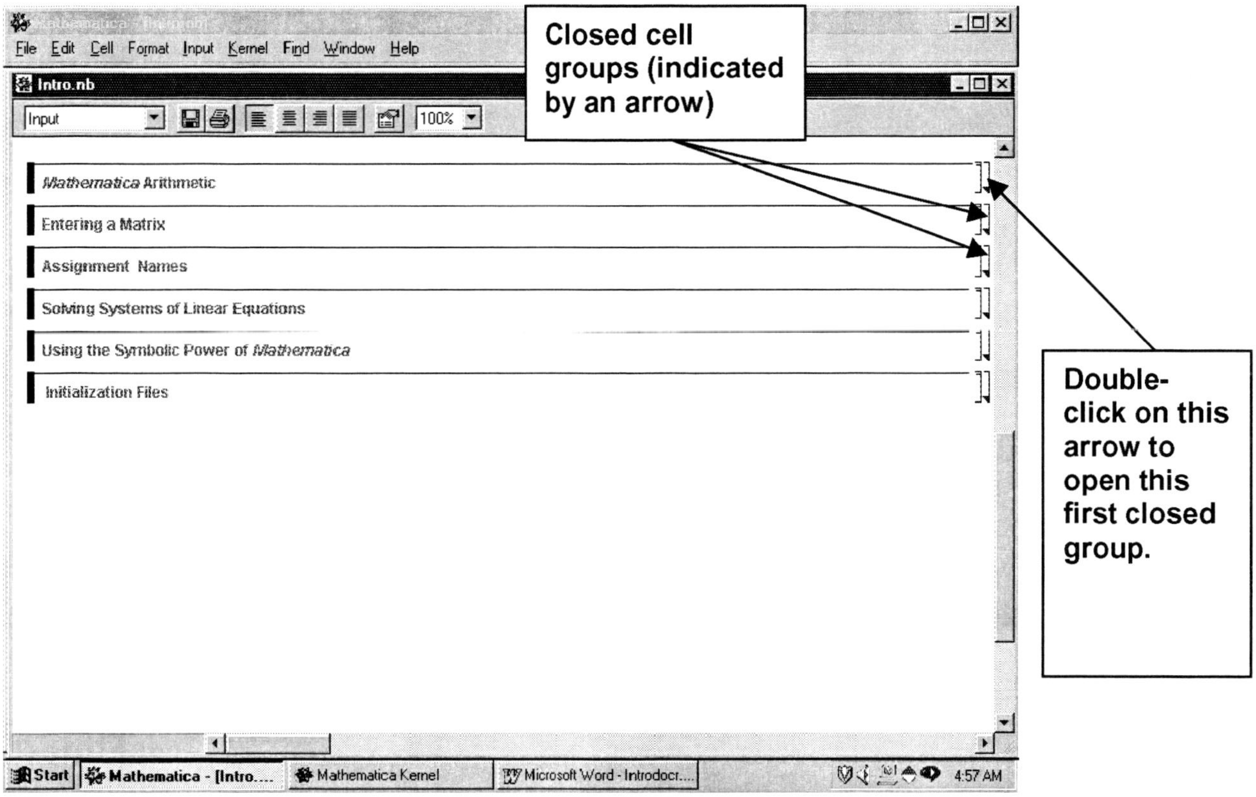

Figure 2: Example of several closed groups of cells

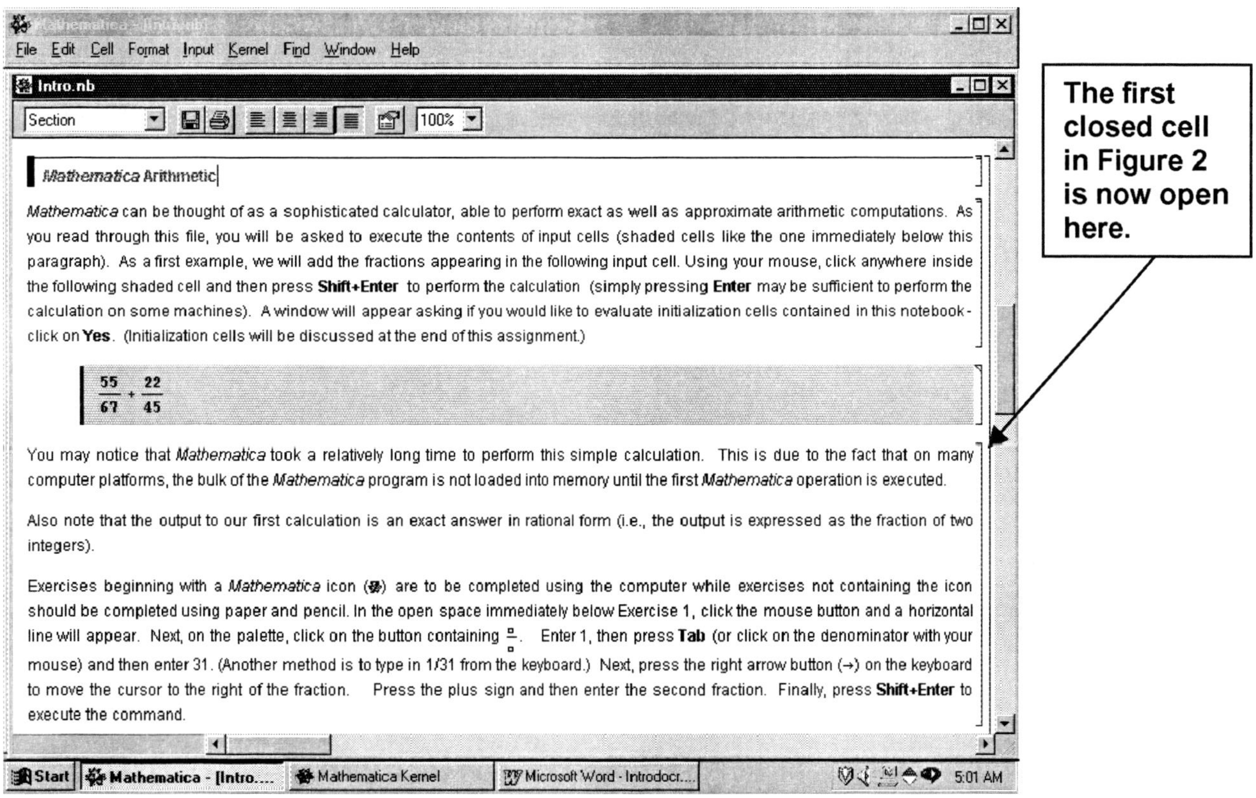

Figure 3: An open cell group

Palettes

One new feature introduced in version 3.0 of Mathematica, which has made Mathematica much easier to use, is the *palette*. Palettes are similar to calculator buttons, providing users with shortcuts to typing in commands from the keyboard. Palettes also extend the keyboard so that other symbols, such as the square root symbol, can be entered into the notebook simply by using the mouse and clicking on the square root button in the palette. Palettes are especially useful to new users of Mathematica who are initially unfamiliar with the syntax of this computer language.

Each computer assignment contains a palette just after the introduction (see Figure 4). In order to activate and use this palette, we must first create a separate window containing this palette. Select the right cell bracket associated with the palette cell as illustrated in Figure 4. Then select File from the menu followed by General Palette from Section as illustrated in Figure 5. A separate window containing the palette will appear (see Figure 6). Drag the window to the right of the screen and resize the notebook containing the assignment so that the assignment and palette appear side-by-side in non-overlapping windows as illustrated in
Figure 7.

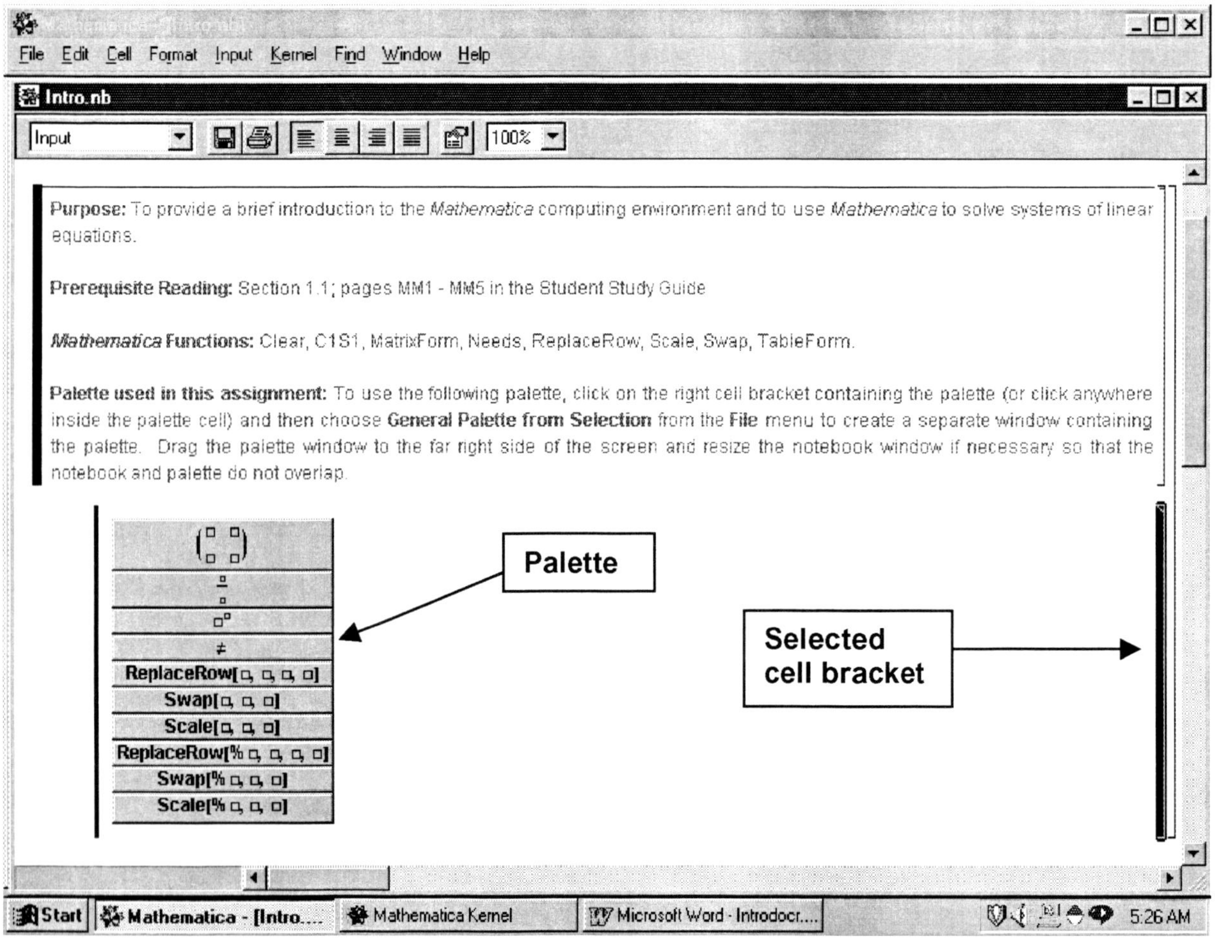

Figure 4: Example of a palette in a computer assignment

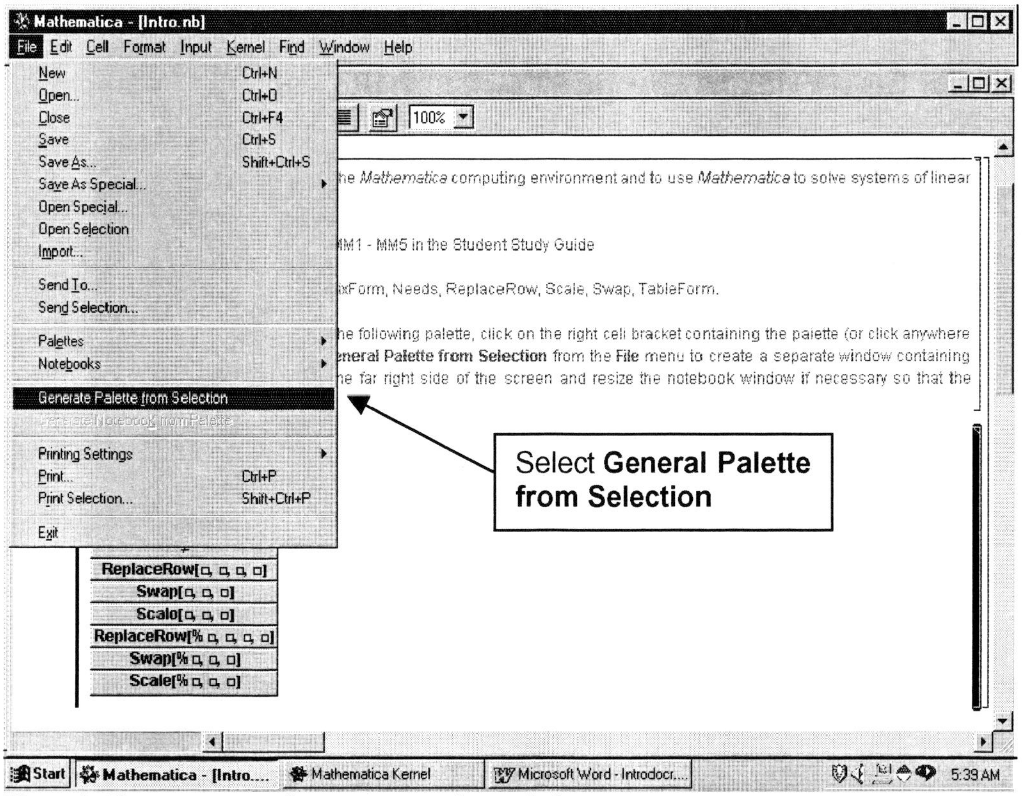

Figure 5: First step to creating a separate palette window

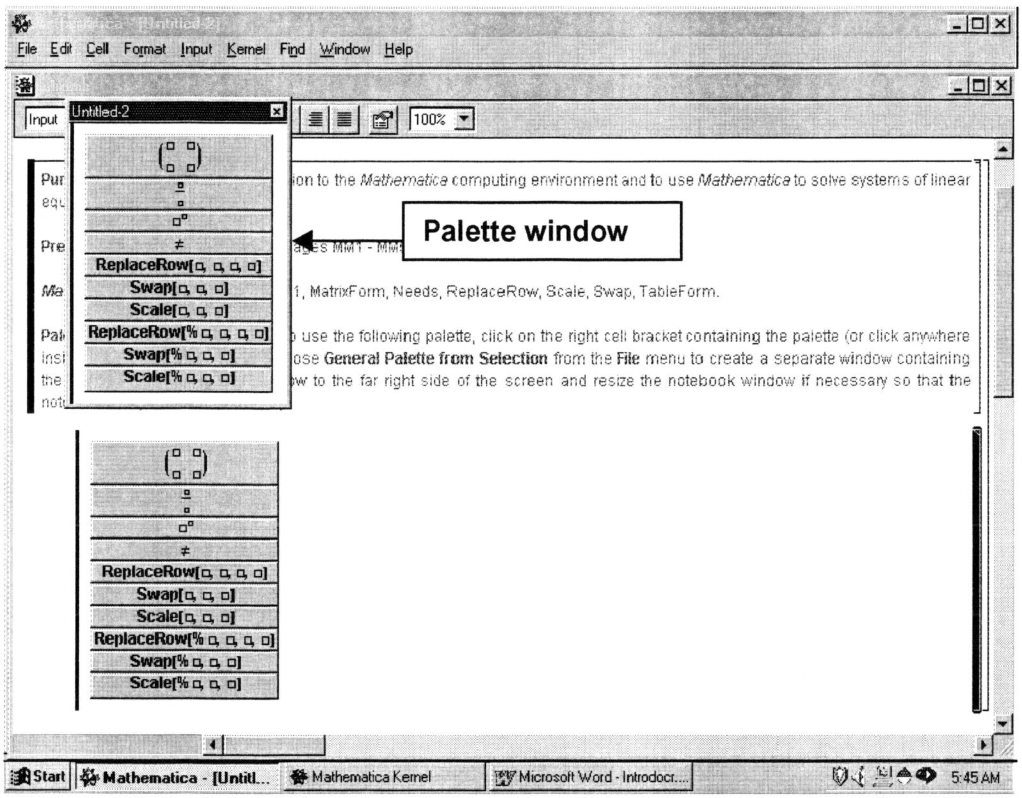

Figure 6: Separate window for the palette appears on the screen

Figure 7: Resized Windows

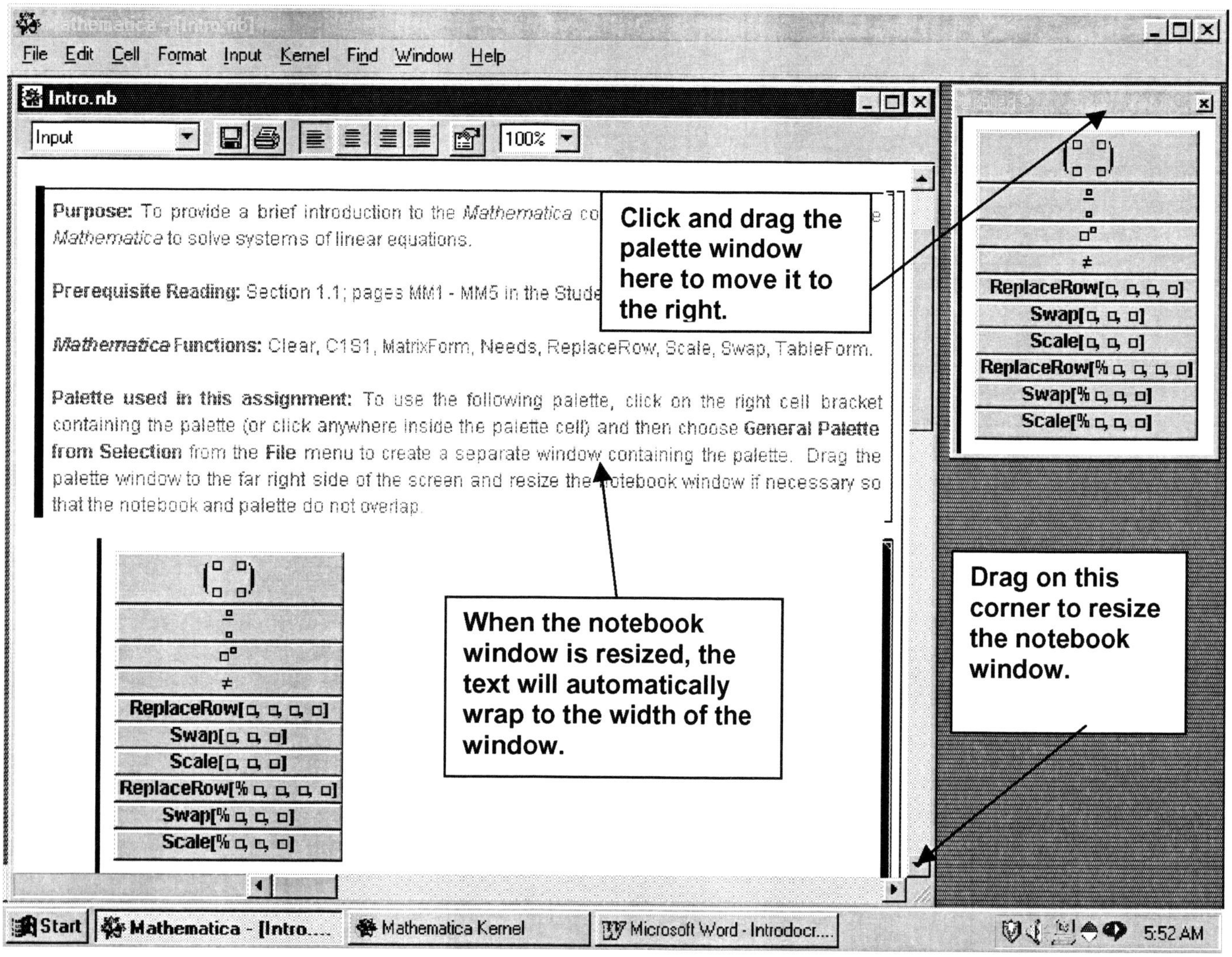

Performing computations in a Notebook

As an example of performing a computation in a notebook, consider Exercise 1 in the first assignment, Introduction to Mathematica. Unless otherwise stated, students should solve Mathematica exercises immediately after the statement of the exercise. To do this, click in the space immediately after the cell containing Exercise 1 and a horizontal line should appear (see Figure 8). Then click on the fraction button in the palette and a fraction will appear in the notebook (see Figure 9). Now enter the number 1, then press the Tab key and then press 31 to complete the first fraction. Now press the right arrow key and then enter the plus sign followed by the second fraction. Press Shift+Enter (on some computer platforms, pressing Enter is sufficient) to perform the calculation (see Figure 10). Notice that the input and output are labeled with In[3] and Out[3]. This implies that 2 other commands were executed prior to this computation.

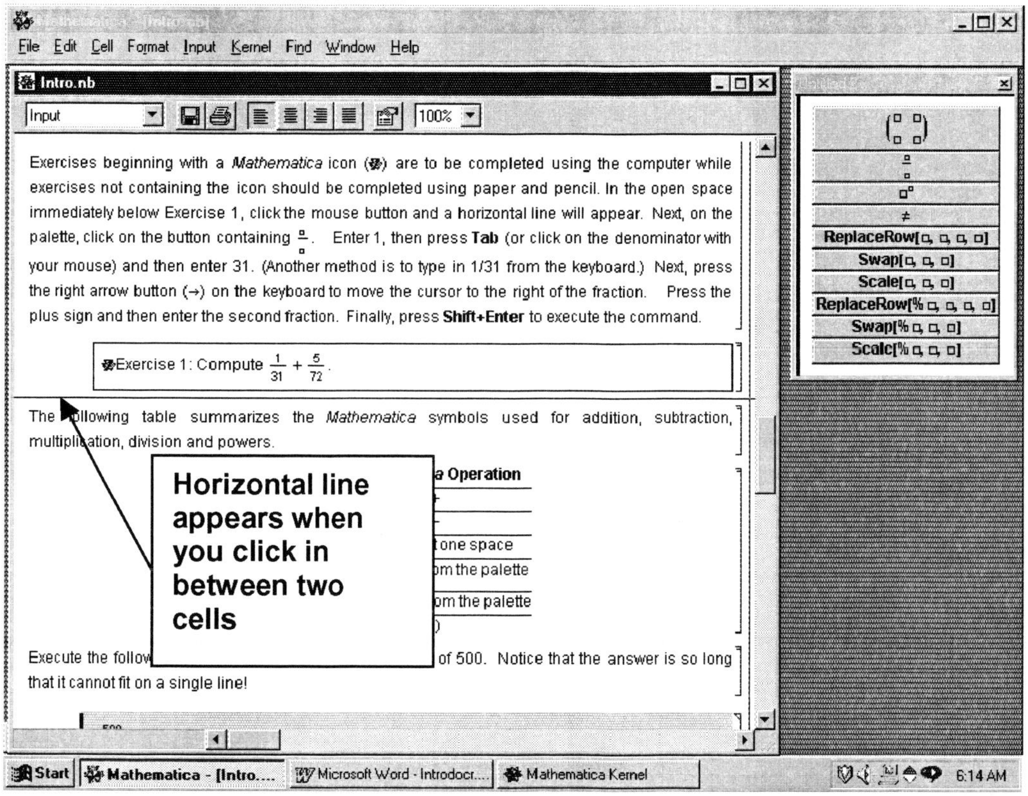

Figure 8: First step to inserting an exercise solution

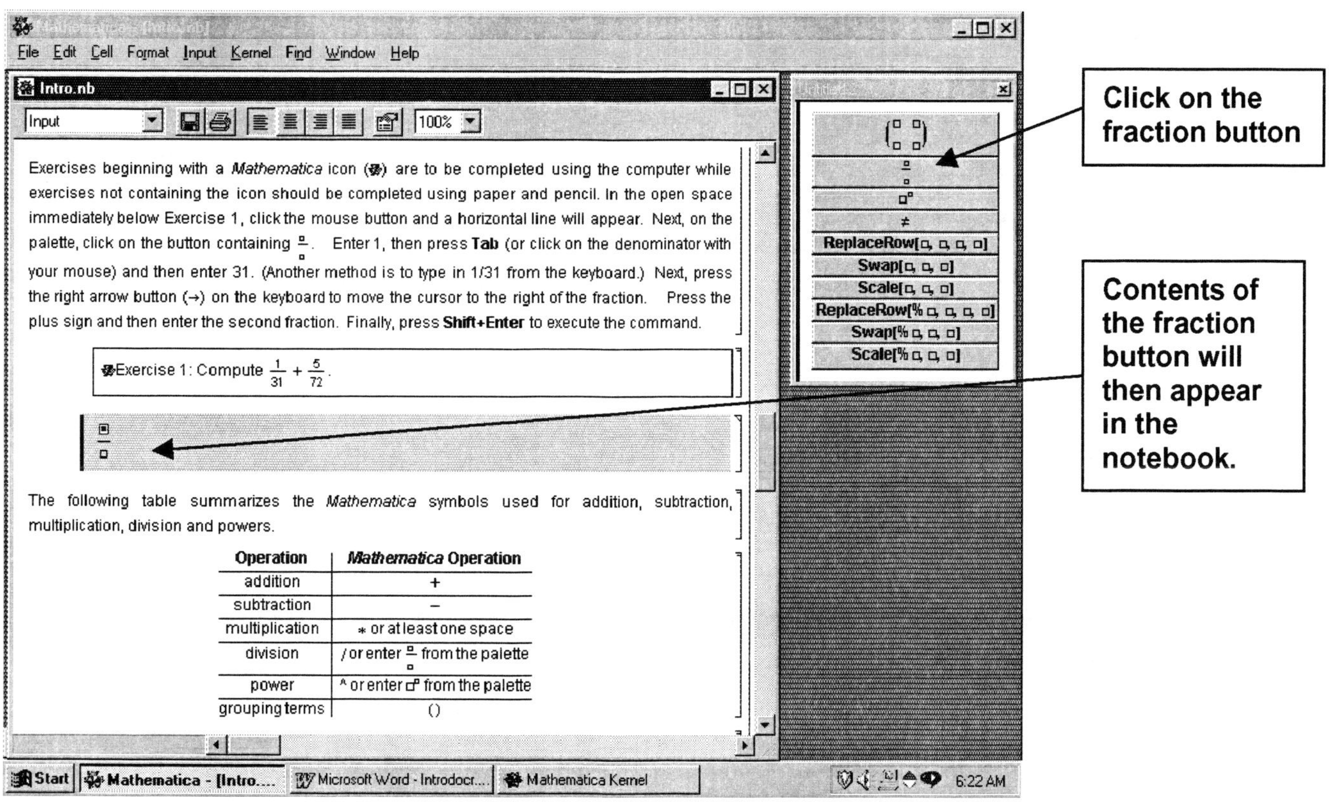

Figure 9: First step to entering a fraction using the palette

Figure 10: Performing an arithmetic computation with Mathematica

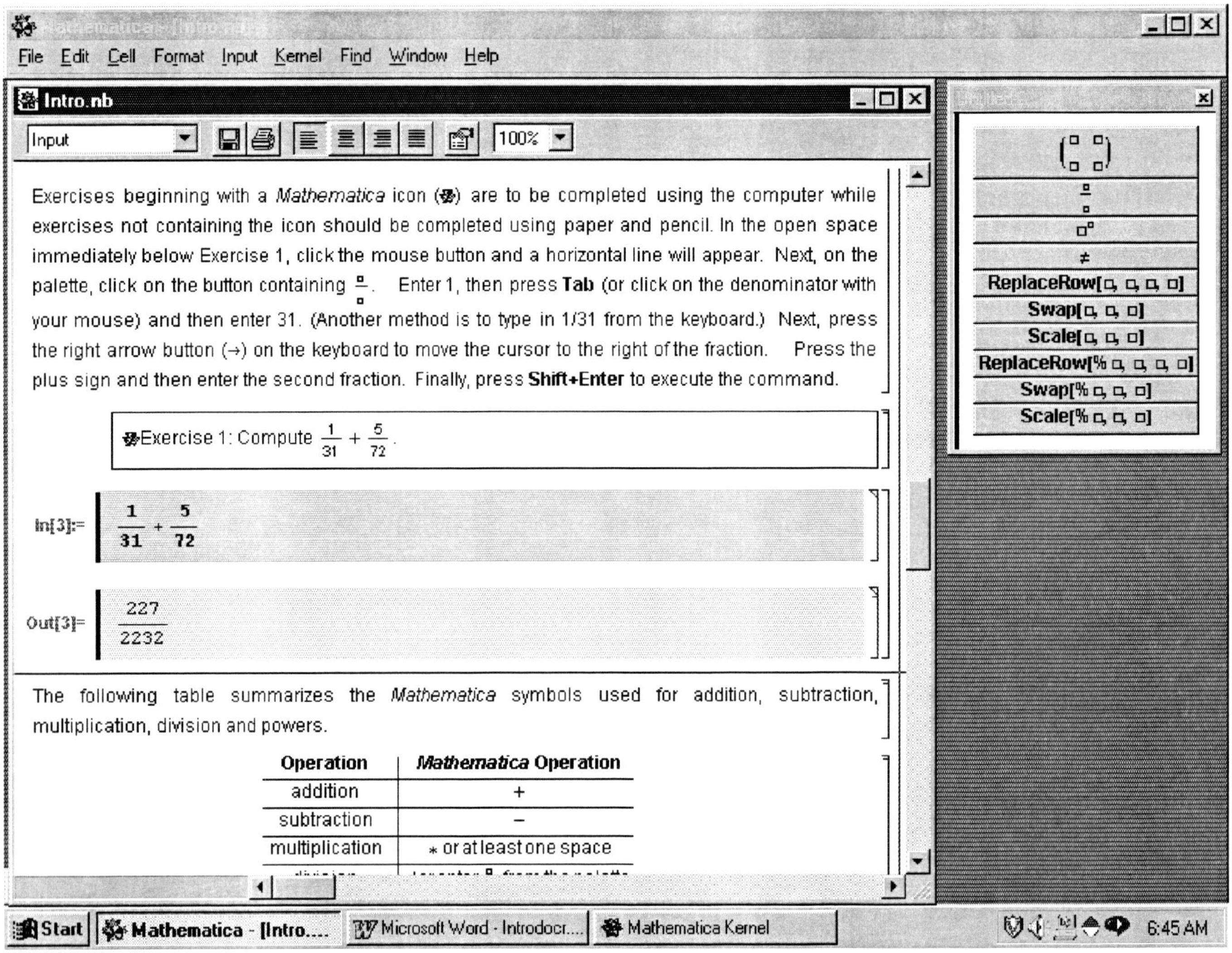

Adding and Deleting Cells

If you plan on plan on revising the electronic notebooks containing computer assignments, then you will need to know how to delete, edit and add cells. To delete a cell, simply click on the cell bracket associated with the cell you want to erase. Then press the Delete key.

To insert a new cell, click in between the two cells where you wish to insert a new cell. By default, the cell will be an input cell if you begin typing at the chosen location. If instead, you want the cell to be of another type, select Format from the menu and select the type of cell you wish to create. Then enter in your new material. You may want to spend some time experimenting with creating new cell types. You may also want to experiment with different font styles and sizes using the Format menu.

Initialization Cells

Most of the computer assignments come with *initialization cells* containing Mathematica commands that will be automatically loaded into memory when the user executes the first command in a computer session. These initialization cells, located at the end of each computer assignment, may contain customized commands created specifically for the computer assignment you are currently working with along with commands to load functions from the Lay Toolbox. The initialization cells will also contain a command so that all matrix output is displayed in matrix form. When a student is working with one of the computer assignment notebooks and executes the first command in a computer session, a message may appear on the screen asking the user if he or she would like to automatically evaluate the initialization cells (see Figure 11). The user should always click Yes.

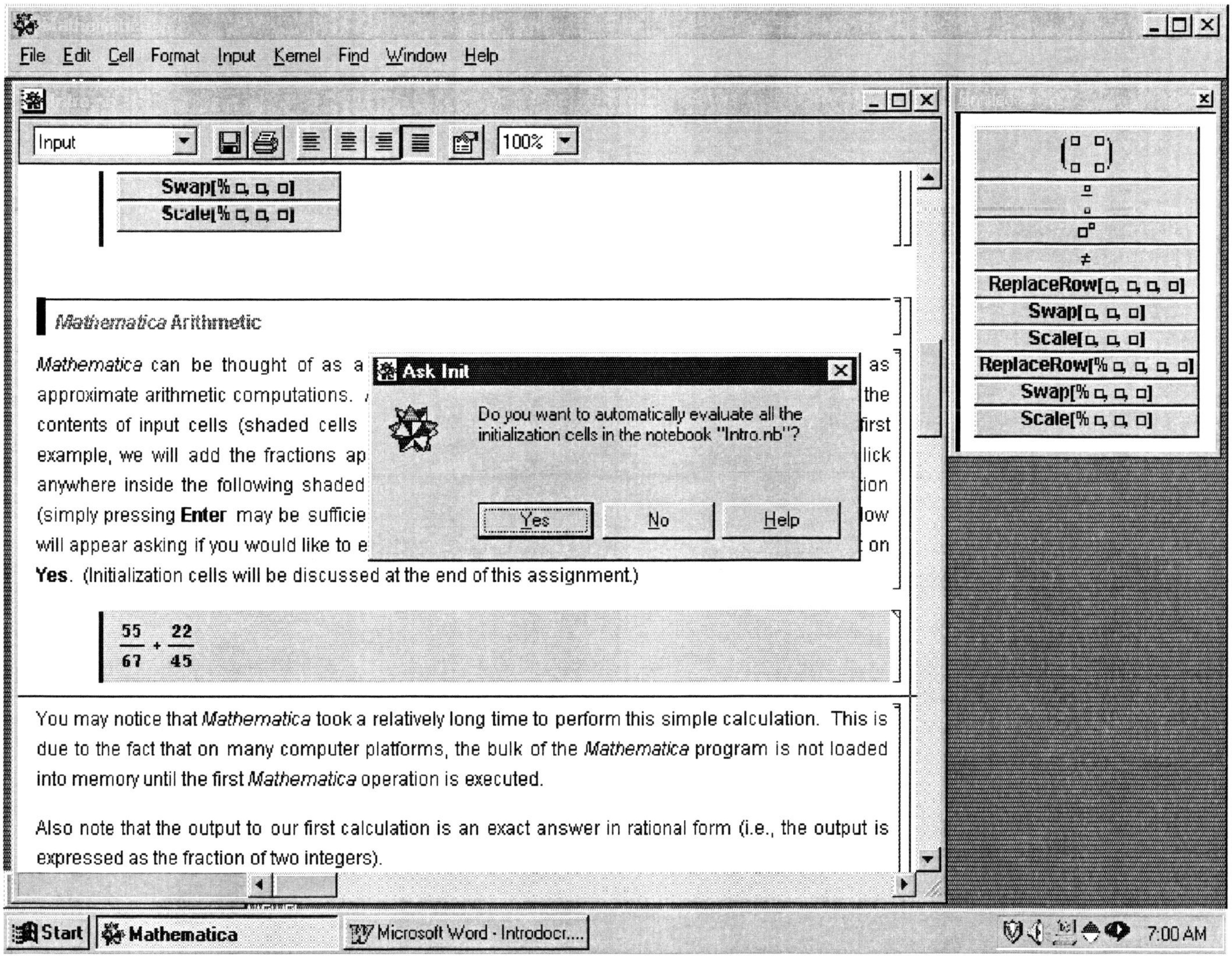

Figure 11: The Ask Init window – click Yes

Add-On Packages

In the introduction to each computer assignment, the purpose, prerequisite reading, and Mathematica functions used in the assignment are declared. Some of the declared functions are standard functions which Mathematica automatically loads into memory at the beginning of a computer session while other functions, contained in add-on packages, are not automatically loaded into memory. Functions in add-on packages are loaded into memory using a Needs["folder`package`"] command. For example, the Lay Toolbox functions are in a package called LayFunctions and this package is contained in a folder called LayData. To load the LayFunctions package, the command Needs["LayData`LayFunctions`"] would need to be executed before using any of the commands contained in this package. Fortunately for you and your students, the computer assignments contained in this manual which use functions in add-on packages, contain appropriate Needs commands in the initialization cells. Therefore these commands will be automatically loaded into memory at the beginning of the computer session. But keep in mind that you will need to use the Needs command when working with [M] exercises (see pages MM-1 and MM-5 in the *Study Guide* for more information).

Some of the computer assignments contain additional Mathematica commands which I wrote specifically for the computer assignments. For example, in the computer assignment entitled Transmitting Secret Codes, there are two commands, MessageToMatrix and MatrixToMessage, which I created to make this assignment easier for students to use. I placed the code for these functions directly in the initialization cells along with some other supporting functions. Since the code for these commands are contained in the initialization cells, these functions will be automatically created at the beginning of a computer session.

Text-based users will probably want to avoid assignments containing more than 4 or 5 initialization cells since all of these commands will have to be entered and executed by the user before beginning the computer assignment.

5. COMMON PROBLEMS STUDENTS ENCOUNTER

FILE STORAGE PROBLEMS

The most prominent problem encountered by students when working with Mathematica is saving and storing notebooks. Students encounter problems storing their notebooks on floppy disks lacking adequate storage space. *Entire Mathematica notebooks can be corrupted and destroyed if a student attempts to save the notebook to a floppy disk lacking adequate storage space.* If possible, students should store their work on a hard drive first and then store a backup of their work on a floppy disk. If hard drive storage is not an option, students should be instructed to save their Mathematica work on a floppy disk containing at least 75% free space. Instruct students to purchase a pack of floppy disks and to format these disks before use.

SYNTAX ERRORS

Mathematica commands are case-sensitive and every standard command begins with a capital letter. If a command is not capitalized, an error message will return when the command is executed or the output will contain undesired results. This will be a more prominent problem for students using the text-based interface since all commands must be typed in from the keyboard. Students using the notebook interface will already have all the necessary commands located in the palette. The use of palettes has almost completely eliminated syntax errors!

BASIC COMPUTER LITERACY ERRORS

Most students will be computer literate and will need very little instruction in opening, saving, closing and printing files. But there will often be a few students who are unfamiliar with the computer environment. Therefore instructors should provide students with clear information about downloading and storing notebooks.

TEXT-BASED USER ERRORS

Instructors and students limited to a text-based interface will need to enter and execute all the commands contained in the initialization cells. Therefore if a command does not work correctly, then make sure that all the initialization cells were correctly entered and executed. Entering and executing a command before it has been properly defined has to be removed from memory using the Remove command before it is defined and used again. For example, suppose you are working with the computer assignment Transmitting Secret Codes and you happen to use the command MessageToMatrix before it has been properly defined. You must first execute the command Remove[MessageToMatrix] and then properly define the command before continuing. Perhaps a better way to avoid this problem is to not use computer assignments containing more than 4 or 5 initialization cells.

STARTUP ERRORS

A common problem your students may encounter when working with Mathematica 3.0 is a startup error as depicted in Figure 12. This is probably a bug in the Mathematica program resulting from the Help menu window being minimized, but not closed, when previous user of Mathematica quit a session with Mathematica. Tell your students to click on the Continue Anyway button if they encounter this error at the beginning of a Mathematica session.

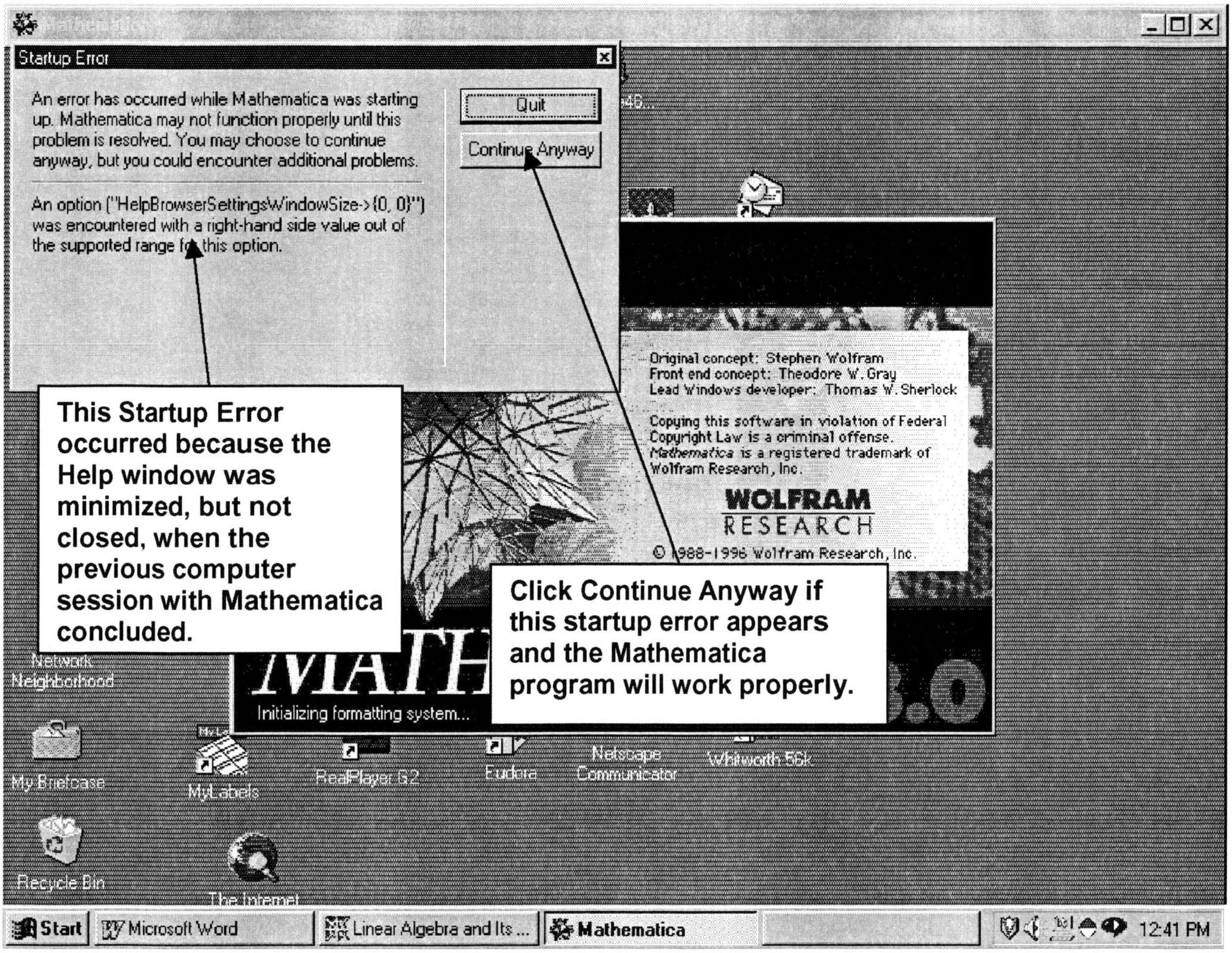

Figure 12: A Startup Error

6. COMPUTER ASSIGNMENTS, CASE STUDIES AND PROJECTS

The first part of this section describes the computer assignments contained in this manual and the remaining part of this section describes the Case Studies and Projects found on the Web/CD.

OVERVIEW OF MATHEMATICA COMPUTER ASSIGNMENTS

In this section, a brief description of each computer assignment is given. Suggestions of what to discuss in class are given in this section along with ways of modifying the assignments to better fit your needs. At least one of the first two assignments must be completed before the other assignments are started and the assignment Computer Graphics must be completed before students attempt to work Motion Pictures. The remaining assignments can be completed independently of the other assignments. At the end of the computer assignment summaries is a table listing the prerequisite reading needed to complete each assignment.

As explained in the first computer assignment, by default the Mathematica program will display all output in list form, not matrix form. For instructors and students using the text-based interface, execute the commands

```
$Post := If[ MatrixQ[#] || VectorQ[#], MatrixForm[#], # ]&
```

and

```
Needs["LayData`LayFunctions`"]
```

at the beginning of every computer session. The first command will instruct the Mathematica program to display all matrix and vector output in matrix form rather than in list form and the second command will load the functions found in the LayFunctions Toolbox. For users of the notebook interface, these commands will automatically be executed when the initialization cells are executed at the beginning of the computer session.

Introduction to Mathematica

In this first assignment, students will learn how to enter and manipulate matrices with Mathematica. Students will also use the commands ReplaceRow, Scale and Swap to solve systems of linear equations.

Interpolating Polynomials

This computer assignment is based upon Section 1.2 and provides students with additional practice with the row reduction algorithm. The last part of the assignment, after exercise 8, introduces a shortcut to curve fitting and informally introduces least-squares problems - you can omit these remaining sections to create a shorter assignment. Be sure to tell students to complete exercises 1 and 2 with paper and pencil before completing the rest of the assignment using a computer. Exercise 11 is a long problem and therefore you may want to assign only part of this exercise or you may wish to delete this exercise from the assignment.

Tracking Rent-a-Cars

Linear difference equations are introduced and are used to analyze the movement of rental cars over a period of time. This assignment is intended in part to be a discovery exercise where students observe the convergence of rental car distributions to a fixed value, even when different initial distributions are used.

Elementary Matrices

This is a discovery exercise intended to teach students some of the basic properties of elementary matrices and to help students understand the justification for the algorithm used to find the inverse of a matrix.

Transmitting Secret Codes

This assignment provides students with an application of matrix multiplication and matrix inversion. Students will use these concepts to encode and decode messages. In the last exercise, students will send an encoded message to the instructor. *This is a popular assignment among the students.* I created two functions specifically for this assignment called MessageToMatrix and MatrixToMessage which require two pages of code (as indicated in the initialization files) and it is therefore this assignment is not recommended for text-based users.

LU Factorization

This assignment begins by reviewing the LU factorization method found on in Section 2.5 in Lay's textbook. Practice using this method is given and then the need for partial pivoting is discussed along with a storage scheme. If you just want to provide your students with further practice finding the LU factorization of a matrix, assign only the first four exercises. The remaining portion of the assignment is a good source for engineering and computer science students who may need to know how the LU factorization of a matrix is usually stored in a single matrix.

Computer Graphics

This assignment (based upon Section 2.8) and the Motion Pictures assignment are the most popular assignments. The last exercise can be omitted if you wish to shorten this assignment.

Motion Pictures

The previous assignment, Computer Graphics, is a prerequisite for this assignment. Students use matrix multiplication and the concepts learned in the previous assignment to create simple computer animations. This assignment makes extensive use of the Table command and it is not recommended for text-based users due to the large number of initialization cells.

Properties of Determinants

This assignment is a set of discovery exercises in which your students will examine some of the important relationships between determinants and elementary row operations. Set a due date for this project *before* beginning chapter 3. Present the definition of a 2x2 determinant in Section 2.2 to your students in class and then tell them that they will see the general definition of a determinant in Chapter 3. The purpose of this assignment is to explore some of the properties of determinants.

Markov Chains

This assignment is based upon Section 4.9 and introduces three methods used to find steady state vectors with Mathematica.

Survival of the Spotted Owls

Eigenvalues and eigenvectors are introduced in this assignment to explore variants of the spotted owl problem introduced at the beginning of Chapter 5 and discussed in Section 5.6.

The following table provides users with a summary of the computer assignments and prerequisite reading.

Computer Assignment	**Electronic File Name**	**Prerequisite Reading in the Textbook**
Introduction to Mathematica	Intro	Section 1.1
Interpolating Polynomials	Poly	Section 1.2
Tracking Rent-a-Cars	Rent	Sections 1.9, 2.1
Elementary Matrices	Elmat	Section 2.2
Transmitting Secret Codes	Code	Sections 2.1 and 2.2
LU Factorization	LUfact	Section 2.5
Computer Graphics	Graphics	Section 2.8
Motion Pictures	Motion	Section 2.8 and completion of the Computer Graphics assignment.
Properties of Determinants	Det	Section 3.1
Markov Chains	Markov	Section 4.9
Survival of the Spotted Owls	Owls	Introduction to Chapter 5, Sections 5.1 and 5.6.

OVERVIEW OF CASE STUDIES

Chapter 1: Linear Models in Economics

This case study examines Leontief's "exchange model" and shows systems of linear equations can model an economy. Real economic data is used.

Chapter 2: Computer Graphics in Automotive Design

This case study explores how a three-dimensional image is rendered effectively in two dimensions. Perspective projections, rotations, and zooming are discussed and applied to wireframe data derived from a 1983 Toyota Corolla.

Chapter 3: Determinants in Analytic Geometry

This case study examines how determinants may be used to find the equations for lines, circles, conic sections, planes, spheres, and quadric surfaces.

Chapter 4: Space Flight and Control Systems

This case study studies a mathematical model for engineering control systems. The notion of rank is used to determine whether a system is controllable, and a system of linear equations is solved to determine which inputs into the system yield a desired output.

Chapter 5: Dynamical Systems and Spotted Owls

This case study examines how eigenvalues and eigenvectors can be used to study the change in a population over time. Real data from populations of spotted owls, blue whales, and plants (speckled alders) is studied, and the notion of a sustainable harvest is introduced.

Chapter 6: Least-Squares Solutions

This case study uses the method of least squares to fit linear, polynomial, and sinusoidal curves to real data. This data includes performance in the Olympic men's 400-meter run, climatic data from Charlotte NC, and tidal data from the Cape Hatteras pier.

Chapter 7: The Singular Value Decomposition and Image Processing

This case study examines how the singular value decomposition of a matrix may be used to reduce the amount of data needed to store a reasonable image of a graphical object. Two types of images are considered: three-dimensional surfaces and black-and-white two-dimensional pictures.

OVERVIEW OF APPLICATION PROJECTS

Section 1.2: Interpolating Polynomials

This set of exercises shows how a system of linear equations may be used to fit a polynomial through a set of data points. Polynomial curves are fit to real data taken from Car and Driver magazine.

Section 1.2 Splines

This set of exercises shows how a system of linear equations may be used to fit a piecewise-polynomial curve through a set of data points. Cubic splines are fit to real data taken from Car and Driver magazine.

Section 1.9 Diet Problems

This set of exercises provides examples of vector equations that result from balancing nutrients in a diet. Real data from the USDA website are used.

Section 1.9 Traffic Flow Problems

This set of exercises shows how a system of linear equations may be used to model the flow of traffic through a network. Real data from the Seattle Transportation Management Division is used in this exploration.

Section 1.9 Loop Currents

This set of exercises provides further examples of loop currents, and reinforces the text's development of this topic.

Section 2.1 Adjacency Matrices

This set of exercises studies the adjacency matrix of a graph. The real route maps of various airlines help to motivate graphical questions which may be answered with adjacency matrices.

Section 2.1 Dominance Matrices

This set of exercises applies matrices to questions concerning competition between individuals and groups. The problem of ordering teams within a football conference is discussed, and real data from the Southeastern Conference, the Big 12 Conference, and the National Football League are used.

Section 2.1 Other Matrix Products

This set of exercises introduces and explores the properties of two matrix products: the Jordan product and the commutator product.

Section 2.3 Condition Numbers

This set of exercises motivates the definition of the condition number of a matrix, and explores how its value affects the accuracy of solutions to a system of linear equations.

Section 2.5 The LU and QR Factorizations

This set of exercises shows how to use an LU factorization to perform a QR factorization. The QR factorization is introduced in Exercise 24 of Section 2.5 in the text.

Section 2.5 Equilibrium Temperature Distributions

This set of exercises discusses the problem of determining the equilibrium temperature of a thin plate. An appropriate system of equations is derived, and is solved both by finding a matrix inverse and by an LU factorization.

Section 2.7 The Leontief Input-Output Model

This set of exercises provides three real data examples of the Leontief Input-Output Model discussed in the text. American economic data from in 1947 and 1992 are studied.

Section 3.3 The Jacobian and Change of Variables

This set of exercises is designed for students who have experienced multivariate calculus. The Jacobian is derived and applied to the change of variables in double and triple integrals.

Section 4.1 Hill Substitution Ciphers

This set of exercises studies how matrices ma be used to encode and decode messages. Matrix arithmetic modulo 26 is used.

Section 4.6 Error-Detecting and Error-Correcting Codes

This set of exercises studies how to construct methods for detecting and correcting errors made in the transmission of encoded messages. The United States Postal Service bar code is studied as an error-detecting code, and the error-correcting Hamming (7,4) code is also studied.

Section 5.3 The Fibonacci Sequence and Generalizations

This set of exercises introduces the Fibonacci sequence and Lucas sequences. Eigenvalues, eigenvectors and diagonalization are used to derive general formulas for an arbitrary element in these sequences.

Section 5.4 Integration by Parts

This set of exercises shows how the matrix of a linear transformation relative to a cleverly chosen basis may be used to find antiderivatives usually found using integration by parts.

Section 6.4 The QR Method for Finding Eigenvalues

This set of exercises shows how the QR factorization of a matrix may be used to calculate the eigenvalues of a matrix. Two methods for performing this action are considered and compared.

Section 6.4 Finding Roots of Polynomials with Eigenvalues

This set of exercises describes how the real roots of a polynomial can be found by finding the eigenvalues of its companion matrix. The QR method is then employed to find these eigenvalues.

Section 7.2 Conic Sections and Quadric Surfaces

This set of exercises shows how quadratic forms and the Principal Axes Theorem may be used to classify conic sections and quadric surfaces.

Section 7.2 Extrema for Functions of Several Variables

This set of exercises is designed for students who have experienced multivariate calculus. Quadratic forms are used to investigate maximum and minimum values of functions of several variables. Results are derived in terms of the eigenvalues of the Hessian matrix.

7. REFERENCES

If you are new to Mathematica, I highly recommend [1] to learn more about Mathematica and its use in undergraduate mathematics. For a short, concise introduction to Mathematica arithmetic, assignment statements, functions and programming structures, see Appendix A in [11]. Electronic versions of [9] and [13], which describe nearly all the capabilities of Mathematica and its standard add-on packages, can be accessed through the Help menu while running the Mathematica program. If you are interested in learning more about using Mathematica specifically for use in a linear algebra course, see [7] and visit the bookstore at www.wolfram.com.

All the projects contained in this lab, except for the project entitled *Transmitting Secret Codes*, are directly based upon material found in Lay's text. For the project *Transmitting Secret Codes*, the ideas for encoding and decoding messages are based upon Jane Day's MATLAB computer project (see [5]). The brief discussion about letter frequencies is based upon information found in [2] and I found [7] to be a very helpful source of ideas for writing some of the functions found in the initialization cells for this project.

Porter's discussion of rotations in [10] was helpful in one of the exercises in *Computer Graphics* assignment. The computer assignment, *LU Factorization*, is based upon information that can be found in a variety of numerical analysis and matrix computation textbooks including [6] and [12].

For more examples of computer exercise (which were written for MATLAB, but could be modified for use with Mathematica), see [8]. For a larger reference list of available resources for teaching linear algebra, see [4].

References

1. Martha L. Abell and James P. Braselton, *Mathematica by Example*, Second Edition, Academic Press, 1997.

2. Albrecht Beutelspacher, *Cryptology*, The Mathematical Association of America, 1994.

3. D. Carlson, C.R. Johnson, D.C. Lay, A.D. Porter, "The Linear Algebra Curriculum Study Group Recommendations for the First Course in Linear Algebra," *College Math. Journal* (24), 1993, 41-46.

4. D. Carlson, C.R. Johnson, D.C. Lay, A.D. Porter, A. Watkins, W. Watkins, eds., *Resources for Teaching Linear Algebra*, MAA Notes No. 42, Math. Assoc. of America, Washington, D.C., 1997.

5. Jane M. Day, *Instructor's MATLAB Manual,* Addison-Wesley, 1997.

6. Gene H. Golub and Charles F. Van Loan, *Matrix Computations,* The Johns Hopkins University Press, 1996.

7. Eugene Johnson, *Linear Algebra with Mathematica,* Brooks/Cole Publishing Company, 1995.

8. Steven Leon, Eugene Herman, and Richard Faulkenberry. ATLAST Computer Exercises for Linear Algebra, Prentice Hall, 1996.

9. Emily Martin ET. Al., Mathematica *3.0: Standard Add-on Packages,* Cambridge University Press, 1996.

10. Gerald J. Porter, "Linear Algebra and Affine Planar Transformations", *College Math. Journal* (24), 1993, 47-51.

11. Robert D. Skeel and Jerry B. Keiper, *Elementary Numerical Computing with* Mathematica, McGraw-Hill, 1993.

12. David S. Watkins, *Fundamentals of Matrix Computations,* John Wiley and Sons, 1991.

13. Stephen Wolfram, *The* Mathematica *Book*, Cambridge University Press, 1996.

8. HARD COPIES OF THE COMPUTER ASSIGNMENTS

The following eleven computer assignments are hard copies of the corresponding Mathematica notebooks. Feel free to modify the notebooks to fit your needs and those of your students.

Computer Assignment: Introduction to Mathematica

Purpose: To provide a brief introduction to the *Mathematica* computing environment and to use *Mathematica* to solve systems of linear equations.

Prerequisite Reading: Section 1.1; pages MM1 - MM5 in the Student *Study Guide*

***Mathematica* Functions:** Clear, C1S1, MatrixForm, Needs, ReplaceRow, Scale, Swap, TableForm.

Palette used in this assignment: To use the following palette, click on the right cell bracket containing the palette (or click anywhere inside the palette cell) and then choose **General Palette from Selection** from the **File** menu to create a separate window containing the palette. Drag the palette window to the far right side of the screen and resize the notebook window if necessary so that the notebook and palette do not overlap.

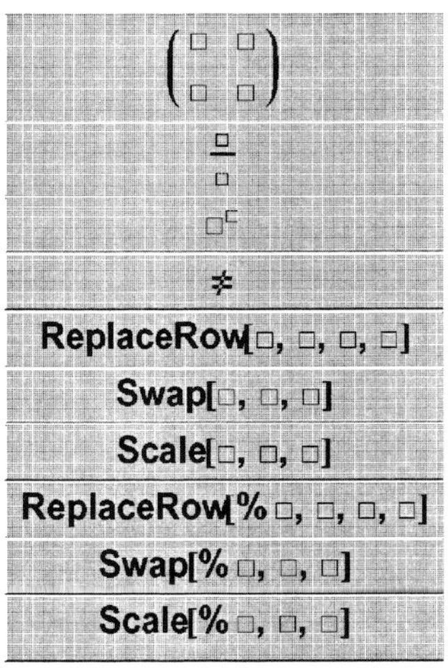

Mathematica Arithmetic

Mathematica can be thought of as a sophisticated calculator, able to perform exact as well as approximate arithmetic computations. As you read through this file, you will be asked to execute the contents of input cells (shaded cells like the one immediately below this paragraph). As a first example, you will add the fractions appearing in the following input cell. Using your mouse, click anywhere inside the following shaded cell and then press **Shift+Enter** to perform the calculation (simply pressing **Enter** may be sufficient to perform the calculation on some machines). A window will appear asking if you would like to evaluate initialization cells contained in this notebook - click on **Yes**. (Initialization cells will be discussed at the end of this assignment.)

$$\frac{55}{67}+\frac{22}{45}$$

You may notice that *Mathematica* took a relatively long time to perform this simple calculation. This is due to the fact that on many computer platforms, the bulk of the *Mathematica* program is not loaded into memory until the first *Mathematica* operation is executed.

Also note that the output to our first calculation is an exact answer in rational form (i.e., the output is expressed as the fraction of two integers).

Exercises beginning with a *Mathematica* icon (✱) are to be completed using the computer while exercises not containing the icon should be completed using paper and pencil. In the open space immediately below Exercise 1, click the mouse button and a horizontal line will appear. Next, on the palette, click on the button containing $\frac{\square}{\square}$. Enter 1, then press **Tab** (or click on the denominator with your mouse) and then enter 31. (Another method is to type in 1/31 from the keyboard.) Next, press the right arrow button (→) on the keyboard to move the cursor to the right of the fraction. Press the plus sign and then enter the second fraction. Finally, press **Shift+Enter** to execute the command.

✱Exercise 1: Compute $\frac{1}{31}+\frac{5}{72}$.

The following table summarizes the *Mathematica* symbols used for addition, subtraction, multiplication, division and powers.

Operation	*Mathematica* Operation
addition	+
subtraction	−
multiplication	∗ or at least one space
division	/ or enter $\frac{\square}{\square}$ from the palette
power	^ or enter $\square^{\square}$ from the palette
grouping terms	( )

Execute the following command to compute 2 to a power of 500. Notice that the answer is so long that it cannot fit on a single line!

```
2^500
```

A space can be used for multiplication. Therefore, to compute 56987 times 23186, at least one space must be inserted between the two numbers. Execute the following computation.

```
56987 23186
```

Parentheses are used to group terms. *Never use square brackets [ ] or curly brackets { } for grouping in arithmetic or algebraic computations. Always use parenthesis ( ).* For example, execute the following arithmetic operation.

```
5 (8 - 6) + 2^5
```

Entering a Matrix

Suppose, for example, you wish to enter the matrix $\begin{pmatrix} 1 & 3 & 5 & 9 \\ 2 & 0 & 3 & 7 \\ -5 & 4 & 3 & 2 \end{pmatrix}$ and assign it to the letter A. One way to create this assignment statement is to enter A={{1,3,5,9},{2,0,3,7},{-5,4,3,2}} where each set of inner brackets represents a row of A. Execute the following command to assign A to the given matrix.

```
A = {{1, 3, 5, 9}, {2, 0, 3, 7}, {-5, 4, 3, 2}}
```

To view A in matrix form, execute the following command:

```
MatrixForm[A]
```

Another way to view a matrix is in **TableForm.** Execute the following command.

```
TableForm[A]
```

Exercise 2: Use brackets { {...},{...},...,{...} } to create the matrix $B = \begin{pmatrix} 1 & 2 & 3 & 4 & 5 \\ 6 & 7 & 8 & 9 & 10 \\ 11 & 1 & 2 & 13 & 0 \end{pmatrix}$.

A more natural way to enter a matrix is to use the palette. Use the mouse and click on the palette button containing $\begin{pmatrix} \square & \square \\ \square & \square \end{pmatrix}$. Then to insert an additional column, press the comma key while holding the **Ctrl** key. Repeat this action to insert additional columns. To insert rows, press the enter key while holding down the **Ctrl** key (i.e. execute **Ctrl+**

Enter). The matrix contained in the following input cell was constructed using this method. Execute the following assignment statement.

$$A = \begin{pmatrix} 1 & 3 & 5 & 9 \\ 2 & 0 & 3 & 7 \\ -5 & 4 & 3 & 2 \end{pmatrix}$$

Once again, the output is not in matrix form. Executing the following command will instruct *Mathematica* to *always* display a matrix in matrix form (you need not understand how this command works, but suffice it to say that before an output statement is displayed, *Mathematica* will check to see if the output is a matrix and if it is, *Mathematica* will display the output in the form of a matrix).

```
$Post := If[MatrixQ[#] || VectorQ[#], MatrixForm[#], #]&
```

Exercise 3: Assign F to equal $\begin{pmatrix} -2.5 & 0 & 2 & 11 \\ 1 & 2 & 3 & 0 \\ -9 & 5 & 7 & -3 \end{pmatrix}$ using the palette. Notice that output will automatically be displayed in matrix form.

Assignment Names

Execute the following assignment statement:

$$D = \begin{pmatrix} 1 & 2 & 3 \\ 1 & 0 & 0 \\ 1 & 0 & 5 \end{pmatrix}$$

When the assignment statement is executed, an error message is displayed indicating that D is a protected symbol. This is due to the fact that **D** is actually a *Mathematica* command used to find the derivative of a function. Other *Mathematica* commands that you have already seen include **MatrixForm** and **TableForm**. *All Mathematica commands are protected symbols and all protected symbols begin with a capital letter.*

So if you create an assignment statement and you obtain a protected symbol error, then you will need to choose a new name in your assignment statement. If you wish to use another capital letter to represent your matrix, avoid using **C**, **D**, **E**, **I**, **N**, and **O** since these are all protected *Mathematica* symbols.

With *Mathematica*, you can use a string of letters for assignment names as long as you begin with a letter of the alphabet and no spaces occur in your assignment statement. An easy way to avoid protected symbol errors is to use an assignment name which begins with a lower case letter. By following this rule of thumb, you will never obtain protected symbol errors since all protected symbols begin with a capital letter. So instead of instead of using the letter **D**, you can use something like **matD**. Execute the following command.

```
matD = (1 2 3
        1 0 0
        1 0 5)
```

Notice that the matrix is displayed again in an output statement. Placing a semicolon at the end of an assignment statement suppresses the display of the output. Execute the following command.

```
matD = (1 2
        0 1);
```

A semicolon can also be used to enter more than one command or assignment statement on a single line. For example, execute the contents of the following cell. The semicolon separates two assignment statements in this case. The semicolon also prevents the output from being displayed.

```
A = (1 2  3
     0 1 -7); B = (-1 3 -2
                    2 5 10);
```

Solving Systems of Linear Equations

Suppose you want to use elementary row operations to solve a system of linear equations:

$$\begin{aligned} 2x_1 + 4x_2 &= 10 \\ 6x_1 + 4x_2 &= 14 \end{aligned}$$

To begin, execute the following command to set up the augmented matrix:

```
M = (2 4 10
     6 4 14)
```

Next you will use the following three *Mathematica* commands to solve the system:

ReplaceRow[M, r, m, s]	Replaces row r of M by row r plus m times row s.
Scale[M, r, c]	Multiplies row r of M by a scalar c.
Swap[M, r, s]	Interchanges rows r and s of M.

The first step in the row reduction of *M* is to replace row 2 by its sum with -3 times row 1. Execute the following command.

```
ReplaceRow[M, 2, -3, 1]
```

Notice that a matrix is now displayed immediately after a statement of the form

Out[n]//MatrixForm=

Determine the specific value of **n** (called the output number) given in this statement. The matrix appearing in this output statement is automatically assigned the name **Out[n]** (**%n** is shorthand for **Out[n]**). For example, if the output number is 19, then the name of the matrix is **%19**.

The next step is to scale row 2 of the matrix just obtained in the last output statement. To do this, place the output number corresponding to the last output statement inside the box (□) in the following scale statement and then execute the command.

```
Scale[% □, 2, -1 / 8]
```

You should obtain an output of $\begin{pmatrix} 2 & 4 & 10 \\ 0 & 1 & 2 \end{pmatrix}$. If not, check to make sure you used the correct output number in the preceding statement. (If not, then reexecute the preceding command after replacing the incorrect output number. To reexecute a command, make the necessary changes to the input cell and then press **Shift+Enter**.) Now place the output number corresponding to this matrix in the box immediately below and then execute the command.

```
ReplaceRow[% □, 1, -4, 2]
```

You should obtain an output of $\begin{pmatrix} 2 & 0 & 2 \\ 0 & 1 & 2 \end{pmatrix}$. If not, check to make sure you used the correct output number in the preceding statement. Now place the output number corresponding to the last output in the following box and then execute the given command.

```
Scale[% □, 1, 1 / 2]
```

The output should be $\begin{pmatrix} 1 & 0 & 1 \\ 0 & 1 & 2 \end{pmatrix}$ and therefore, $x_1 = 1$ and $x_2 = 2$.

Exercise 4:

Solve the following linear system:

$$\begin{array}{rcrcr} & & -2\,x_2 & = & 6 \\ 2\,x_1 & + & 4\,x_2 & = & -8 \end{array}$$

Follow the instructions below to begin this exercise. Once you have found the solution, fill in the missing values immediately below (to enter a number, click inside each empty box (□) and type in the appropriate number).

$x_1 = \square$, $x_2 = \square$

Check your solution (by hand) by substituting the values of x_1 and x_2 into each of the original equations.

Execute the following assignment statement to set up the corresponding augmented matrix.

$$M = \begin{pmatrix} 0 & -2 & 6 \\ 2 & 4 & -8 \end{pmatrix};$$

The first step is to switch rows, by executing the following command:

```
Swap[M, 1, 2]
```

Now continue the process to find a solution.

Exercise 5:

Solve the following linear system:

$$\begin{aligned} 3x_1 - 2x_2 + 6x_3 &= 33 \\ 6x_1 - 4x_2 + 5x_3 &= 31 \\ 9x_1 \qquad + x_3 &= 14 \end{aligned}$$

using the commands **ReplaceRow, Scale** and **Swap.**

Begin by executing the assignment statement below to set up the corresponding augmented matrix. Once you have found the solution, fill in the missing values immediately below:

$x_1 = \square$, $x_2 = \square$, $x_3 = \square$

```
M = (3  -2  6  33
     6  -4  5  31
     9   0  1  14);
```

The next problem is Exercise 20 in Section 1.1.

Exercise 6:

Use the commands **ReplaceRow**, **Scale** and **Swap** to solve the linear system in Exercise 20 of Section 1.1 in the text. Follow the instructions below. See pages MM-5 and MM-6 in the *Study Guide* for more details. Once you have found the solution, fill in the missing values immediately below.

$x_1 = \square$, $x_2 = \square$, $x_3 = \square$, $x_4 = \square$

Execute the following command to load all the exercise data for Chapter 1:

```
Needs["LayData`Chapter1`"]
```

Get the data for Exercise 20:

```
C1S1[20]
```

The matrix displayed is the augmented matrix of the system of equations in Exercise 20. The response suggests that you assign this data to the letter M. So, execute the next command.

```
M = C1S1[20]
```

(Whenever you anticipate that the data will be a matrix, you can skip the command **C1S1[20]**, for example, and just enter **M=C1S1[20]**.) Now that you have M defined, use the row operations from the palette to find the desired solution.

Using the Symbolic Power of *Mathematica*

Suppose you want to determine the values of h such that $\begin{pmatrix} 1 & h & 2 \\ 2 & -4 & 5 \end{pmatrix}$ is the augmented matrix of a consistent linear system. Execute the following cell. (The **Clear** command is used to clear out any value that may have been assigned to *h* during the current computer session.)

```
Clear[h]; M = (1  h  2
               2  -4  5);
```

Now execute the following command:

```
ReplaceRow[M, 2, -2, 1]
```

From observing the bottom row, the system is inconsistent if and only if $-4-2h=0$. So, the system is consistent if and only if $h \neq -2$.

Exercise 7: Complete Exercise 22 on page 11. Begin by executing the following assignment statement and **Clear** command. When you have found the answer, complete the following statement by typing in the remaining part of the sentence inside the box □.

The system is consistent if and only if □.

```
Clear[h]; M = (1  4  -2
               3  h  -6);
```

Exercise 8:

Complete Exercise 26 on page 11. Begin by executing the following cell and when you have found the answer, complete the following statement by replacing each box with an appropriate number.

The system is consistent if and only if □ g + □ h + □ k = □

```
Clear[g, h, k]; M = ( 2   5  -3  g
                      4   7  -4  h
                     -6  -3   1  k);
```

Initialization Cells

At the end of each computer assignment will be a section called **Initialization Cells.** This section contains commands that will be automatically loaded into memory when the first *Mathematica* command is executed. For example, the command **$Post:=If[MatrixQ[#]||VectorQ[#],MatrixForm[#],#]&** introduced earlier will usually be placed in the initialization files so that all matrix output is automatically placed in matrix form. The initialization files will also load any additional functions, needed to complete the given computer assignment, into memory. Some of the computer assignments will also contain functions from Lay's Toolbox along with come customized commands created by the author of these computer assignments. The goal is to allow you to spend your time concentrating on learning linear algebra and a minimal amount of time focusing on *Mathematica* syntax.

The following Needs commands was used to load the functions contained in Lay's Toolbox into memory at the beginning of this computer session. You need not execute this command.

```
Needs["LayData`LayFunctions`"]
```

Computer Assignment: Interpolating Polynomials

Purpose: For a given set of data points, find a polynomial whose graph passes through every point in the data set.

Prerequisites: Read Section 1.2 and read about **Row Operations** on page MM-6 in the *Study Guide*. Then read the paragraph preceding Exercise 35 in Section 1.2 and study the definition of an interpolating polynomial. Complete the first two exercises in this computer assignment before entering the computer lab to do the remaining exercises.

***Mathematica* Functions:** Clear, Fit, ListPlot, Plot, ReplaceRow, Scale, Show, Swap.

Palette used in this assignment: To use the following palette, click on the right cell bracket containing the palette (or click anywhere inside the palette cell) and then choose **General Palette from Selection** from the **File** menu to create a separate window containing the palette. Drag the palette window to the far right side of the screen and resize the notebook window if necessary so that the notebook and palette do not overlap.

```
Clear[p]
p[t_] =
q[t_] =
p[□]
q[□]
Plot[p[t], {t, □, □}]
ListPlot[□, PlotStyle →
 {PointSize[.02],
 RGBColor[□, □, □]}]
Show[□, □]
```

$$\begin{pmatrix} \square & \square \\ \square & \square \end{pmatrix}$$

$$\frac{\square}{\square}$$

$$\square^{\square}$$

```
ReplaceRow[% □, □, □, □]
Scale[% □, □, □]
Swap[% □, □, □]
Fit[□, {□}, t]
```

Plotting Points and Polynomials

Note that exercises containing the *Mathematica* icon (❋) are to be completed using *Mathematica* while the remaining exercises are to be completed using paper and pencil.

Exercise 1: Complete Exercise 35 in Section 1.2 on a separate sheet of paper. Be sure to show your system of equations and the corresponding augmented matrix. Compare your answer with the one listed in the back of the textbook.

Suppose the answer to Exercise 35 was not listed in the back of the book and you wanted to verify that the polynomial you found does indeed pass through the points (1, 12), (2, 15) and (3, 16). Begin by executing the following input cell. *Note that this input cell has a shaded background so that it is easier to see.* To execute the input cell, click anywhere inside the shaded region and then press **Ctrl+Enter** (on some machines, pressing **Enter** is sufficient). A window will appear asking if you would like to evaluate initialization cells contained in this notebook - click on **Yes**. (Initialization cells will be discussed at the end of this assignment.)

```
Clear[p];
p[t_] = 7 + 6 t - t^2
```

The input cell that you just executed contains a **Clear** command and an assignment statement. The semicolon is used to separate these two commands.

The **Clear** command erases any previously defined functions that were assigned to p. The assignment statement always looks strange to new users of *Mathematica. It is essential to place an underscore (_) immediately following the t on left-hand side of the assignment statement and the* underscore *should not appear anywhere else in the assignment statement.* Think of the underscore as a way of declaring *t* to be a variable. Notice also the use of the square brackets. In *Mathematica,* square brackets [...] are used exclusively with functions.

A space was also placed between the 6 and the t. *A space between two numbers, two variables or a number and a variable always indicates multiplication.*

Since the graph of the polynomial passes through the points (1, 12), (2, 15) and (3, 16), you should find that $p(1) = 12$, $p(2) = 15$ and $p(3) = 16$. Execute the following three commands to verify that this is indeed true.

```
p[1]
```

```
p[2]
```

```
p[3]
```

Plotting Points

Plotting a function is easy with *Mathematica*. The command **Plot[p[t], { t, a, b }]** will display the graph of $y = p[t]$ for values of t ranging from $t = a$ to $t = b$. Execute the following statement to plot the polynomial and to assign the name polygraph to the polynomial. (There is nothing special about the name *polygraph* - any name of your own choosing would work as long as the name begins with a lowercase letter and no spaces occur in your chosen name.)

```
polygraph = Plot[p[t], {t, 0, 4}]
```

To plot the points, first form a list of ordered pairs and arbitrarily assign a name to the list. Notice that an ordered pair (x, y) is denoted by **{x, y}**. Then use the command **ListPlot** to display a graph of the points. Execute the following commands.

```
pts = {{1, 12}, {2, 15}, {3, 16}};
ListPlot[pts]
```

You may find it difficult to see the plotted points, so in order to see the points better, the **PlotStyle** option **PointSize[.02]** is used to increase the size of the points and **RGBColor[0,0, 1]** is used to display the points in the color blue. (Note that **RBGColor[1, 0, 0]** represents the color red and **RGBColor[0, 1, 0]** represents the color green.) Execute the following statement and notice that the graph is assigned the name of *ptgraph*.

```
ptgraph = ListPlot[pts, PlotStyle -> {PointSize[.02], RGBColor[0, 0, 1]}]
```

To show the graph of the function and points simultaneously, execute the following **Show** command.

```
Show[ptgraph, polygraph]
```

Exercise 2: Complete this exercise on a separate sheet of paper without the help of *Mathematica*. Find the polynomial whose graph passes through $(-1, 6)$, $(1, 4)$ and $(2, 12)$. Be sure to show your system of equations and the corresponding augmented matrix. You will get a chance to verify your solution in Exercise 3.

Exercise 3: To check your work to Exercise 2 with *Mathematica*, complete the following steps.

(a) Click in the open space immediately below this exercise cell and a horizontal line will appear. This will be the location of your work to this exercise. *Unless otherwise stated, always do your work immediately below the cell containing the statement of the problem.* Click on the **Clear[p]** command in the palette and execute this command to clear out the current definition of function **p**.

(b) Click on **p[t_]=** in the palette and then type in the polynomial found in Exercise 2. Leave a space between the coefficients and the variables to indicate multiplication. Then execute this assignment statement.

(c) Click on **p[□]**, type in **-1** and then execute this statement to compute the value of $p(-1)$. Compute the values of p(1) and p(2) in a similar manner. If the values found are incorrect, then correct your work to Exercise 2 and then redo this exercise. *Note: To redo this exercise, simply click in the cell containing the **Clear[p]** command, that you already created, and reexecute the command. Then edit the cell containing **p[t_]=...** and reexecute this cell. Finally reexecute the cells containing **p[-1]**, **p[1]** and **p[2]**.*

Exercise 4: The goal of this exercise is to plot the points and the polynomial simultaneously. Complete the following:

(a) Create a list containing the data points in Exercise 2 and assign a name to the list (be sure that the list is contained in curly brackets **{...}** and each ordered pair (t, y) is entered in the form **{*t*, *y*}**).

(b) Graph the data points and assign a name of your own choosing to the data points.

(c) Graph the polynomial and assign a name to the graph.

(d) Use the Show command to plot the polynomial and the data points simultaneously.

Performing Row Reduction with *Mathematica*

Using elementary row operations to row reduce a matrix by hand can be slow and tedious. In this section, we will speed up the row reduction process by using the commands **ReplaceRow**, **Scale** and **Swap**. If you need a reminder of how these commands work, read page MM-6 in the *Study Guide.*

Suppose you want to find the polynomial $p(t) = a_0 + a_1\, t + a_2\, t^2$ passing through (1, 8), (2, 25) and (3, 56). After setting up a system of 3 equations and 3 unknowns, you would then find the associated augmented matrix to be the one appearing in the following cell. Execute the assignments statement below.

$$M = \begin{pmatrix} 1 & 1 & 1 & 8 \\ 1 & 2 & 4 & 25 \\ 1 & 3 & 9 & 56 \end{pmatrix}$$

Now we will solve the system of equations by applying elementary row operations to the augmented matrix. Execute the following command.

```
ReplaceRow[M, 2, -1, 1]
```

Recall that **%n** represents the content of output statement **Out[n]**. Execute the following six commands to continue the row reduction process, after first placing the output number of the previous step in each successive box. Study the outcome of each command.

```
ReplaceRow[%□, 3, -1, 1]
```

```
ReplaceRow[%□, 3, -2, 2]
```

```
Scale[%□, 3, 1/2]
```

```
ReplaceRow[%□, 2, -3, 3]
```

```
ReplaceRow[%□, 1, -1, 3]
```

```
ReplaceRow[%□, 1, -1, 2]
```

If you executed the commands correctly, the last output should be in reduced echelon form and the solution should lead to an interpolating polynomial $p(t) = 5 - 4\,t + 7\,t^2$.

Exercise 5: Find the polynomial $p(t) = a_0 + a_1\,t + a_2\,t^2 + a_3\,t^3$ passing through (0,1), (1,-1), (2,-17) and (3,-65) by completing the following steps.

(a) Fill in the missing boxes below to set up a system of linear equations that can be solved to find the values of the unknown coefficients. (To fill in an empty box, simplify click inside the box and enter the appropriate number. When you finish entering the number, press the **Tab** key to go to the next box.)

$a_0 + \square\, a_1 + \square\, a_2 + \square\, a_3 = \square$
$a_0 + \square\, a_1 + \square\, a_2 + \square\, a_3 = \square$
$a_0 + \square\, a_1 + \square\, a_2 + \square\, a_3 = \square$
$a_0 + \square\, a_1 + \square\, a_2 + \square\, a_3 = \square$

(b) Set up an augmented matrix corresponding to the system of equations found in part a. Begin by typing in **M=** and then click on $\begin{pmatrix} \square & \square \\ \square & \square \end{pmatrix}$ in the palette. Use **Ctrl-Enter** to insert extra rows and press the comma key while holding down the hold **Ctrl** key to insert extra columns. Execute the assignment statement.

(c) Use *Mathematica* to row reduce the corresponding augmented matrix. When using the elementary row operation commands, enter all constants as fractions of two integers in order to maintain exact, fractional entries in your output. For example, instead of entering a command like

ReplaceRow[%n,1,0.5,2], enter **ReplaceRow[%n,1,$\frac{1}{2}$,2]**. (To enter the fraction $\frac{1}{2}$, use the palette button containing $\frac{\square}{\square}$ or enter 1/2.)

Exercise 6: Redefine the function p to be the polynomial you found in the previous exercise, after first clearing out the current definition of p. Then check your work to Exercise 5 by evaluating $p(t)$ for each of the given values of t in Exercise 5. *Be sure to use square brackets as demonstrated earlier! Go back and correct Exercise 5 if you find any errors and then redo this exercise.*

Exercise 7: Plot the points and polynomial simultaneously. See your work to Exercise 4 for review if necessary.

Exercise 8: Attempt to find the polynomial $p(t) = a_0 + a_1 t + a_2 t^2$ passing through (0, 1), (1, −1), (2, −17) and (3, −65) by first filling in the boxes below.

$a_0 + \square a_1 + \square a_2 = \square$
$a_0 + \square a_1 + \square a_2 = \square$
$a_0 + \square a_1 + \square a_2 = \square$
$a_0 + \square a_1 + \square a_2 = \square$

Then use *Mathematica* to row reduce the corresponding augmented matrix. What happens?

Write the description of your answer here:

The Fit Command and Least Square Polynomials

Another quicker way of finding a polynomial $p(t) = a_0 + a_1 t + \ldots + a_{n-1} t^{n-1}$ passing through the points (t_1, y_1), (t_2, y_2), ..., (t_n, y_n) is to use the **Fit** command. For example, to find the polynomial of degree 2 passing through (0, 3), (2, 7) and (−2, 5), begin by executing the following command to create a list containing these points.

```
data = {{0, 3}, {2, 7}, {-2, 5}}
```

To find the polynomial, we use the command **Fit[data, { 1, *t*, *t*² }, *t*]** where the last t in the command indicates that the polynomial is to be defined in terms of t and the list $\{1, t, t^2\}$ contains the powers of t to be found in the polynomial. Execute the following command to find the polynomial and to assign it to the function $p[t]$.

```
Clear[p];
p[t_] = Fit[data, {1, t, t^2}, t]
```

Exercise 9: Use the fit command to find the polynomial whose graph passes through $(-1, 6)$, $(1, 4)$ and $(2, 12)$. Verify your work by comparing your results to Exercise 3.

Approximating Polynomials

Since the graph of a polynomial of degree 1 is a straight line, there is no such line passing through $(-1, 6)$, $(1, 4)$ and $(2, 12)$. But the following commands can be used to find the line which lies "closest" to the three points. (A precise definition of "closest" is given in Section 6.5.) Execute the following commands and study the output of each statement.

```
Clear[p];
p[t_] = Fit[data, {1, t}, t]
```

```
dataplot = ListPlot[data, PlotStyle -> {PointSize[.02], RGBColor[0, 0, 1]}]
```

```
lineplot = Plot[p[t], {t, -2, 2}]
```

```
Show[lineplot, dataplot]
```

Exercise 10: Use the Fit command to find a polynomial of the form $p(t) = a_0 + a_1 t + a_2 t^2$ passing "closest" to the points $(0, 1)$, $(1, -1)$, $(2, -17)$ and $(3, -65)$. Plot the points and the polynomials simultaneously.

Exercise 11: Study Exercise 36 on page 26. In this exercise, let *t* represent the velocity and let *p*[*t*] represent force.

a. Complete the exercise by setting up a system of six equations and six unknowns and then row reducing the corresponding augmented matrix. Be sure to show all of your work on a separate sheet of paper. Check your work using the **Fit** command.

b. Define p to be the polynomial you found in part a and compute p[750] to estimate the force when the velocity is 750 ft/sec.

c. Try setting up and solving a system of 4 equations and 4 unknowns in an attempt to find a polynomial of degree 4 whose graph passes through the given points. What happens?

d. Use the Fit command to find a polynomial of degree 4 whose graph lies *closest* to data points. Define q to be this polynomial and compute q[750] to estimate the force when the velocity is 750 ft/

sec.

e. Plot the graph of the data points and polynomial q simultaneously.

Initialization Cells

At the end of each computer assignment will be a section called **Initialization Files.** This section contains commands that will be automatically loaded into memory when the first *Mathematica* command is executed. For example, the command **$Post:=If[MatrixQ[#]||VectorQ[#],MatrixForm[#],#]&** introduced earlier will usually be placed in the initialization files so that all matrix output is automatically placed in matrix form. The initialization files will also load any additional functions, needed to complete the given computer assignment, into memory. Some of the computer assignments will also contain functions from Lay's Toolbox along with come customized commands created by the author of these computer assignments. The goal is to allow you to spend your time concentrating on learning linear algebra and a minimal amount of time focusing on *Mathematica* syntax.

The following Needs commands was used to load the functions contained in Lay's Toolbox into memory at the beginning of this computer session. You need not execute this command.

```
Needs["LayData`LayFunctions`"];
```

```
$Post := If[MatrixQ[#] || VectorQ[#], MatrixForm[#], #]&;
```

Computer Assignment: Tracking Rent-a-Cars

Purpose: To use matrix multiplication to study linear difference equations.

Prerequisite Reading: Read the section entitled **Difference Equations** on pages 90-92 in your textbook. You will also need to be familiar with matrix multiplication introduced in Section 2.1. In the *Study Guide*, read about the **Representation of a Vector** on page MM-7, **Multiplying a Matrix and a Vector** on page MM-8, **Generating a Sequence** on page MM-9, and **Matrix Operations** starting on the bottom of page MM-9.

***Mathematica* Functions:** Clear, Table, ListPlot, MatrixPower, Factor, Round.

Palette used in this assignment:

```
C1S9[□]
Table[□, {□, □, □}]
x[□]
x[0] =
x[k_] :=
MatrixPower[□, □]
x[0] = {r₀, s₀}
Clear[x]
// Factor
// Round
```

Tracking Budget Rent A Cars

> Exercise 1: The data for Section 1.9 in your textbook will be automatically loaded into memory when you execute your first *Mathematica* command in this computer session. Use the command **C1S9** to load the data for Exercise 12 in Section 1.9. Then create an assignment statement to assign *M* and ***x*0** to the given matrix and vector, respectively.

To view the matrix and vector in matrix form, execute the following commands:

```
M
```

```
x0
```

The objective of Exercise 12 is to find $\mathbf{x}_2$, which is not too difficult to compute by hand. But computing $\mathbf{x}_k$ with paper and pencil for larger values of k would be tedious and time-consuming. To quickly obtain values of $\mathbf{x}_k$ with *Mathematica*, a function **x** satisfying **x[k]** = $\mathbf{x}_k$ for $k = 0, 1, 2, 3, \ldots$ is constructed in the following input cell. The **Clear** command clears out any previously defined values of **x** and the assignment statement **x[0] = x0** assigns the value of $\mathbf{x}_0$ to **x[0]** . By the definition of a difference equation, it follows that $\mathbf{x}_k = M\mathbf{x}_{k-1}$ for $k = 1, 2, 3$, and so forth. This recursive relation is expressed as **x[k_]** := *M*.**x[k – 1]** with *Mathematica*. The underscore immediately following k on the left hand side of the assignment statement is necessary in order to define k to be a variable and the colon-equals sign (:=), which is a *delayed* equals sign, must be used instead of an equals sign (=), when defining **x** recursively. Execute the following:

```
Clear[x];
x[0] = x0;
x[k_] := M . x[k - 1]
```

Execute the following command to obtain the value of $\mathbf{x}_2$, the approximate number of cars at each location on Wednesday.

```
x[2]
```

Exercise 2: Determine the number of cars at each location on Thursday and then on Saturday. Do your work immediately below this exercise cell and then type in your answers below by filling in the missing boxes, rounding your answers to the nearest whole numbers.

Number of cars at each location:

On Thursday:	Airport : □	On Saturday:	Airport : □
	East : □		East : □
	West : □		West : □

The command **Table[x[k] , {k, 1, n}]** generates and displays the values of **x[1]**, **x[2]**, **...** , **x[n]**. For example, to observe the distribution of cars over the next 75 days, execute the following command and note that output is in matrix form. The kth *row* of the output corresponds to **x[k]**.

```
Table[x[k], {k, 1, 75}]
```

To round each number in the table to the nearest whole number, execute the following command instead.

```
Table[x[k], {k, 1, 75}] // Round
```

Exercise 3: Study the output you just obtained using the **Table** command.

(a) Determine the number of cars at each location after 75 days by observing the value of ***x*[75]** (round each answer to the nearest whole number):

Airport: □

East: □

West: □

(b) At first, the number of cars at each location changes each day. Does this pattern of daily redistribution continue for all 75 days? (Explain)

Type in your answer here:

Suppose that Budget Rent A Car officials are concerned about the small number of cars at the east side office. To increase the number of cars at the east side office, the number of cars are redistributed so that there are 75 cars at the airport, 300 cars at the east side office and 75 cars at the west side office. Execute the following command.

```
x[0] = {75, 300, 75}
```

Exercise 4: Complete each part below.

a. Use the **Table** command, along with **//Round,** to display the values of ***x*[1]**, ***x*[2]**, ..., ***x*[75]**.

b. Determine the number of cars at each location after 75 days by observing the value of ***x*[75]** (round your answers to the nearest whole numbers):

Airport: □

East: □

West: □

c. Compare the number of cars at each location after 75 days with Exercise 3. (Type in your answer below.)

State your answer here:

d. Redistributing the cars in this exercise certainly increased the number of cars at the east side office initially, but are there more cars at the east side office after 75 days compared to Exercise 3? (Type in your answer below.)

State your answer here:

Exercise 5: Redefine ***x*[0]** to be a redistribution of the cars of your own choosing. Make sure that the total number of cars in your redistribution is 450. Combine the **Table** and **//Round** commands to display the distribution of cars **x[1]** through **x[75]**. Compare your results on the 75th day with the previous two exercises and comment.

State your comments here:

Exercise 6: Based upon your work to the previous exercises, state a conjecture about the eventual number of cars at each location after several weeks have passed.

Airport: □ East: □ West: □

Predicting Long Range Rent A Car Distributions

Exercise 7: This exercise and Exercise 8 are designed to help you predict the eventual number of cars at each of the three locations for any number of total cars and for any distribution. Be sure to review **Powers of a Matrix** on page 106 in your textbook first. You have been working with a difference equations of the form $\mathbf{x}_k = M\mathbf{x}_{k-1}$. Hence $\mathbf{x}_1 = M\mathbf{x}_0$. Fill in the seven blank boxes (□) in the equations below (you may want to do your work on scratch paper first, before filling the boxes).

$$\mathbf{x}_2 = M\mathbf{x}_1 = M(\square\, \mathbf{x}_0) = M^{\square}\, \mathbf{x}_0$$

$$\mathbf{x}_3 = M\mathbf{x}_2 = M(\square\, \mathbf{x}_1) = M^{\square}\, \mathbf{x}_0$$

$$\mathbf{x}_4 = M\mathbf{x}_3 = M(\square\, \mathbf{x}_2) = M^{\square}\, \mathbf{x}_0$$

$$\text{In general, } \mathbf{x}_k = M^{\square}\, \mathbf{x}_0$$

The *Mathematica* command **MatrixPower[M,k]** will compute M^k. By your answer to the last part of Exercise 6, another way to compute **x[k]** is to compute **MatrixPower[M, k] . x[0]**.

Suppose $\mathbf{x}_0 = \begin{pmatrix} a_0 \\ e_0 \\ w_0 \end{pmatrix}$ so that a_0, e_0, and w_0 represent the number of cars at the airport, the east side office and the west side office, respectively. Execute the following cell to clear out the current definition of x and to redefine **x[0]** and **x[k]**.

```
Clear[x];
x[0] = {a₀, e₀, w₀};
x[k_] := MatrixPower[M, k] . x[0]
```

Execute the following command below.

```
x[175]
```

To factor the result, add **//Factor** to the end of the statement. Execute the following command.

```
x[175] // Factor
```

Exercise 8:

Determine the values of ***x*[200]** and ***x*[300]**, factoring the results in each case.

Exercise 9: Answer the following questions *without* making any further computations with *Mathematica*.

a. What does the sum $a_0 + e_0 + w_0$ represent?

State your answer here:

b. For any total number of cars Budget Rent A Car has distributed among the three locations, what will be the eventual percentage cars at each location?

State your answer here:

c. Suppose Budget Rent A Car has a total of 1000 cars distributed among the three locations. What will be the eventual number of cars located at each location?

State your answer here:

Predicting Long Range Population Distributions

Exercise 10:

(a) Read Example 3 in Section 1.9 and use the command **C1S9** to load the data corresponding to Exercise 13. Use the suggested names listed in the output to create an appropriate assignment statement.

(b) Define ***x*[0]** to be the value of ***x*0**.

(c) Use the **MatrixPower** command to define ***x*[*k*]**.

(d) Determine the population distribution in 2010 and 2050.

(e) Redefine ***x*[0]** to equal $\{r_0, s_0\}$. (Hint: Use the palette.) Then determine the long-term distribution of the percentage of total people living in the city and and in the suburbs. Fill in the following boxes (□):

Percentage of people living in the :

City: □

Suburbs: □

Initialization Cells

```
$Post := If[VectorQ[#] || MatrixQ[#], MatrixForm[#], #]&
```

```
Needs["LayData`Chapter1`"]
```

Computer Assignment: Elementary Matrices

Purpose: To understand the relationship between elementary matrices and the inverse of a given nonsingular matrix.

Prerequisite Reading: Read Section 2.2 in the textbook and read about the commands **IdentityMatrix** and **Inverse** on page MM-10 in the *Study Guide*.

***Mathematica* Functions:** Clear, IdentityMatrix, Inverse, ReplaceRow, Scale, Swap

Palette Used in this Assignment:

$$\begin{pmatrix} \square & \square \\ \square & \square \end{pmatrix}$$

```
ReplaceRow[□, □, □, □]
Swap[□, □, □]
Scale[□, □, □]
Inverse[□]
```

Elementary Matrices and Matrix Multiplication

Execute the following **Clear** command.

```
Clear[a, b, c, d, e, f, g, h, i, j, k, l, m, n, o, p]
```

Execute the contents of the following cell. (The semicolons are used to separate the given assignment statements and to prevent output from being displayed.)

```
A = {{a,b,c,d},{e,f,g,h},{i,j,k,l},{m,n,o,p}}; E1 = {{1,0,0,0},{5,1,0,0},{0,0,1,0},{0,0,0,1}}; E2 = {{1,0,0,0},{0,1,0,0},{-5,0,1,0},{0,0,0,1}}; E3 = {{1,0,0,0},{0,1,0,0},{0,0,1,0},{0,2,0,1}};
E4 = {{1,0,0,7},{0,1,0,0},{0,0,1,0},{0,0,0,1}};
```

$$A = \begin{pmatrix} a & b & c & d \\ e & f & g & h \\ i & j & k & l \\ m & n & o & p \end{pmatrix};\ E1 = \begin{pmatrix} 1 & 0 & 0 & 0 \\ 5 & 1 & 0 & 0 \\ 0 & 0 & 1 & 0 \\ 0 & 0 & 0 & 1 \end{pmatrix};\ E2 = \begin{pmatrix} 1 & 0 & 0 & 0 \\ 0 & 1 & 0 & 0 \\ -5 & 0 & 1 & 0 \\ 0 & 0 & 0 & 1 \end{pmatrix};\ E3 = \begin{pmatrix} 1 & 0 & 0 & 0 \\ 0 & 1 & 0 & 0 \\ 0 & 0 & 1 & 0 \\ 0 & 2 & 0 & 1 \end{pmatrix};$$

$$E4 = \begin{pmatrix} 1 & 0 & 0 & 7 \\ 0 & 1 & 0 & 0 \\ 0 & 0 & 1 & 0 \\ 0 & 0 & 0 & 1 \end{pmatrix};$$

Matrices, such as E1 through E4, are called *elementary matrices* (see page 113 in your textbook).

Execute the following product (recall that the period is used for matrix multiplication).

```
E1.A
```

Examine the output above and note that the performing this product is the same as replacing row 2 of *A* by row 2 plus with 5 times row 1. Therefore the product E1.*A* equals **ReplaceRow[A, 2, 5, 1]**. (To recall the definition of **ReplaceRow,** see page MM-6 in the *Study Guide*.) Execute the following command to verify this result.

```
ReplaceRow[A, 2, 5, 1]
```

Exercise 1: Study the matrix E2 and compute E2.*A*. Then execute a **ReplaceRow** command which produces the same result as E2.*A*.

Exercise 2: Study the matrix E3 and compute E3.*A*. Then execute a **ReplaceRow** command that accomplishes the same task as E3.*A*.

Exercise 3: Study the matrix E4 and then compute E4.*A*. Execute a **ReplaceRow** command which produces the same result as E4.*A*.

Exercise 4: For the following two input cells, execute the **ReplaceRow** command and then fill in the missing entries for E5 so that E5.*A* produces the same result as **ReplaceRow[A, 3, -5, 2].** Then execute the second input cell to verify your work.

```
ReplaceRow[A, 3, -5, 2]
```

```
E5 = (1 □ □ □
      □ 1 □ □
      □ □ 1 □
      □ □ □ 1); E5 . A
```

Exercise 5: For the following two input cells, execute the **ReplaceRow** command and then fill in the missing entries for E6 so that E6.*A* produces the same result as **ReplaceRow[A,2,9,4].** Then execute the second input cell to verify your work.

```
ReplaceRow[A, 2, 9, 4]
```

```
E6 = (1 □ □ □
      □ 1 □ □
      □ □ 1 □
      □ □ □ 1); E6 . A
```

Execute the following input cell.

```
E1 = (3 0 0 0       E2 = (1 0  0 0       E3 = (0 0 1 0
      0 1 0 0             0 1  0 0             0 1 0 0
      0 0 1 0  ;          0 0 -5 0  ;          1 0 0 0  ;
      0 0 0 1)            0 0  0 1)            0 0 0 1)
```

If necessary, review the definitions of **Swap** and **Scale** on page MM-6 in the *Study Guide*.

Exercise 6: Study matrix E1 and compute E1.*A*. Then execute a **Scale** command that produces the same matrix as E1.*A*.

Exercise 7: Study matrix E2 and compute E2.*A*. Then execute a **Scale** command that produces the same matrix as E2.*A*.

Exercise 8: Execute the **Scale** command below. Then fill in the missing entries in the assignment statement for E5 so that E5.*A* equals **Scale[A,2,5]**. Execute the product to verify your work.

```
Scale[A, 2, 5]
```

```
     ⎛□ 0 0 0⎞
E5 = ⎜0 □ 0 0⎟; E5 . A
     ⎜0 0 □ 0⎟
     ⎝0 0 0 □⎠
```

Exercise 9: Examine the contents of matrix E3. (Notice that E3 results from switching rows 1 and 3 of the identity matrix.) Execute the product E3 . *A* and then execute a **Swap** command producing the same result as E3 . *A*.

Exercise 10: Create a matrix E5 so that E5 . *A* has the effect of switching rows 2 and 4 of matrix *A*. Verify your work by computing E6 . *A*.

Row Reduction through Matrix Multiplication

Execute the following assignment statement for *A*.

```
    ⎛ 1   2   3⎞
A = ⎜-2  -4  -8⎟;
    ⎝ 3   8   1⎠
```

It can be shown that *A* is non-singular. Suppose you wanted to row reduce the matrix *A* to the identity matrix *I*. A possible first step is to replace row 2 of *A* by 2 times row 1 plus row 2. Execute the following input cell and make sure you understand what is happening.

```
     ⎛1 0 0⎞
E1 = ⎜2 1 0⎟; E1 . A
     ⎝0 0 1⎠
```

Matrix E2 is used to introduce a zero in the first column of row 3 of the last output. Execute the following cell after first filling in the empty box with the output number associated with the last output.

```
     ⎛ 1 0 0⎞
E2 = ⎜ 0 1 0⎟; E2 . %□
     ⎝-3 0 1⎠
```

Continue the row-reduction process by executing the rest of the assignment statements and products below in order. Be sure to enter the output number of the preceding product before computing each product.

```
E3 = {{1, 0, 0}, {0, -1/2, 0}, {0, 0, 1}}; E3 . % □
```

$$E3 = \begin{pmatrix} 1 & 0 & 0 \\ 0 & \frac{-1}{2} & 0 \\ 0 & 0 & 1 \end{pmatrix};\ E3\ .\ \%$$

$$E4 = \begin{pmatrix} 1 & 0 & 0 \\ 0 & 1 & 0 \\ 0 & 0 & \frac{1}{2} \end{pmatrix};\ E4\ .\ \%$$

$$E5 = \begin{pmatrix} 1 & 0 & 0 \\ 0 & 0 & 1 \\ 0 & 1 & 0 \end{pmatrix};\ E5\ .\ \%$$

$$E6 = \begin{pmatrix} 1 & 0 & 0 \\ 0 & 1 & 4 \\ 0 & 0 & 1 \end{pmatrix};\ E6\ .\ \%$$

$$E7 = \begin{pmatrix} 1 & -2 & 0 \\ 0 & 1 & 0 \\ 0 & 0 & 1 \end{pmatrix};\ E7\ .\ \%$$

$$E8 = \begin{pmatrix} 1 & 0 & -3 \\ 0 & 1 & 0 \\ 0 & 0 & 1 \end{pmatrix};\ E8\ .\ \%$$

The last output should be the identity matrix.

Reviewing the steps that led to the matrix *I* just obtained, the following 8 products were computed:

1: E1 . *A*

2: E2 . (E1 . *A*)

3: E3 . (E2 . E1 . *A*)

and so forth all the way to the last product

8: E8 . (E7 . E6 . E5 . E4 . E3 . E2 . E1 . *A*)

Since E8 . (E7 . E6 . E5 . E4 . E3 . E2 . E1 . *A*) = (E8 . E7 . E6 . E5 . E4 . E3 . E2 . E1) . *A* , the complete row reduction process can be completed in one step by computing (E8 . E7 . E6 . E5 . E4 . E3 . E2 . E1) . *A*.

Exercise 11: Before executing the contents of the following cell, write down what you expect the product to equal. Why? Type in your answer and reasoning below.

Answer:

```
(E8 . E7 . E6 . E5 . E4 . E3 . E2 . E1) . A
```

Exercise 12: Why is (E8.E7.E6.E5.E4.E3.E2.E1) the inverse of *A*? Type in your answer below.

Answer:

Exercise 13: You have seen that applying the elementary row operations, represented by E1 through E8, to matrix A produces the identity matrix. Suppose instead that these elementary row operations are applied to the identity matrix *I*. That is, if we compute (E8.E7.E6.E5.E4.E3.E2.E1).*I*, then what do we obtain? (Fill in the empty box below.)

Fill in: The □ is obtained by computing (E8.E7.E6.E5.E4.E3.E2.E1).*I*

Why? (Explain here.)

Exercise 14: Explain why the same elementary row operations used to row reduce a nonsingular matrix *A* to the identify matrix *I* also transforms *I* into the inverse of *A*. Write your explanation below:

Explanation:

Exercise 15: Execute the following assignment statement and then find E1, E2, ..., En so that the product En...E2.E1.*A* equals the identity matrix (verify your work with *Mathematica*). Then use the elementary matrices to find the inverse of *A*. Verify your work using the **Inverse** function.

$$A = \begin{pmatrix} 0 & 1 & 0 \\ 1 & 5 & 0 \\ 2 & 3 & 1 \end{pmatrix};$$

Initialization Cell

```
Needs["LayData`LayFunctions`"];
$Post := If[MatrixQ[#] || VectorQ[#], MatrixForm[#], #]&
```

Computer Assignment: Transmitting Secret Codes

Purpose: To use matrix multiplication and matrix inversion to encode and decode messages.

Prerequisite Reading: Sections 2.1 and 2.2

***Mathematica* Functions:** Inverse, MatrixToMessage, MessageToMatrix, Mod

Palette Used in this Assignment:

```
Inverse[□,
  Modulus -> 29]
MatrixToMessage[□]
MessageToMatrix[□]
Mod[□, □]
(□ □)
(□ □)
```

Introduction: Before sensitive information is passed from one individual to another, it is usually encoded to prevent any unauthorized individuals from understanding the information being sent. Only those individuals who understand the precise encryption process being used will be able to decipher the message.

In this assignment, each letter or character in a message will be substituted with a number to disguise the message and matrix multiplication will be used to further scramble the message. The inverse of the matrix used to scramble the message will be applied to decode the secret message.

Prerequisite Information: Modular Arithmetic

Modular arithmetic will be used in this assignment and hence a brief introduction to this topic is covered here.

It is said that 8 equals 2 modulo 6, written $8 = 2 \pmod 6$, because if you begin counting clockwise at 1 around the following circle of numbers, you will arrive at the number 2 when a count of eight is reached.

```
   1
0     2
5     3
   4
```

Exercise 1: Use the method just described to complete each of the following modular equations. (Click inside each box and insert your answer.)

a. 13 = □ (mod 6) b. 17 = □ (mod 6) c. 22 = □ (mod 6)

Another way to compute the values found in Exercise 1 is to divide each number on the left by 6 and determine the remainder. For example, 13 divided by 6 is 2 with a remainder of 1. So 13 = 1 (mod 6).

Exercise 2: Use the remainder technique just described to verify your answers to Exercise 1, parts b and c (you need not show any work). Then complete each of the following equations. (Click inside each box and insert your answer.) You will verify your answers in Exercise 4.

a. 46=□ (mod 7) b. 56=□ (mod41)

Modular arithmetic can also be performed on matrices. For example, $\begin{pmatrix} 7 & 91 \\ 89 & 74 \end{pmatrix} = \begin{pmatrix} 7 & 1 \\ 9 & 4 \end{pmatrix}$ (mod 10)

since 7 = 7 (mod 10), 91 = 1 (mod 10), 89 = 9 (mod 10) and 74 = 4 (mod 10).

Exercise 3: Complete each of the following equations by hand and insert your answers in the empty boxes provided below. (You will verify your answers in Exercise 5).

a. $\begin{pmatrix} 11 & 31 \\ 15 & 9 \end{pmatrix} = \begin{pmatrix} \square & \square \\ \square & \square \end{pmatrix}$ (mod 7)

b. $\begin{pmatrix} 1 & 17 \\ 0 & 9 \end{pmatrix}\begin{pmatrix} 1 & 1 \\ 0 & 3 \end{pmatrix} = \begin{pmatrix} \square & \square \\ \square & \square \end{pmatrix}$ (mod 26) *(Hint: Multiply the matrices on the left first.)*

The *Mathematica* command, **Mod[m, n]**, gives the remainder on division of m by n. So to verify that 8 = 2 (mod 6), execute the following command.

```
Mod[8, 6]
```

Exercise 4: Use the **Mod** command to verify your answers for parts a and b in Exercise 2. Be sure to correct your work to Exercise 2 if mistakes are found.

Modular arithmetic can also be performed on matrices. For example, to verify that $\begin{pmatrix} 7 & 91 \\ 89 & 74 \end{pmatrix} = \begin{pmatrix} 7 & 1 \\ 9 & 4 \end{pmatrix}$ (mod 10), execute the following command.

```
Mod[(7  91
     89 74), 10]
```

Suppose you want to verify that $\begin{pmatrix} 3 & 1 \\ 5 & 2 \end{pmatrix}\begin{pmatrix} 0 & 2 & 3 \\ 4 & 5 & 6 \end{pmatrix} = \begin{pmatrix} 4 & 1 & 0 \\ 3 & 0 & 2 \end{pmatrix}$ (mod 5). Execute the following **Mod** command.

$$\text{Mod}\left[\begin{pmatrix} 3 & 1 \\ 5 & 2 \end{pmatrix} . \begin{pmatrix} 0 & 2 & 3 \\ 4 & 5 & 6 \end{pmatrix}, 5\right]$$

Exercise 5: Use the **Mod** command to verify your answers in Exercise 3.

Introduction to Cryptography

Cryptography is the process of decoding or encoding messages. Suppose you wish to send the following secret message to a friend so that no one else can read the message.

MEET TONIGHT

You could use the following *Numerical Conversion Table* to encode the message:

A	B	C	D	E	F	G	H	I	J
1	2	3	4	5	6	7	8	9	10
K	L	M	N	O	P	Q	R	S	T
11	12	13	14	15	16	17	18	19	20
U	V	W	X	Y	Z	SPACE	.	?	□
21	22	23	24	25	26	27	28	29	□

Numerical Conversion Table

Replacing each character with the number immediately below it, the following encoded message is obtained:

M	E	E	T	□	T	O	N	I	G	H	T
13	5	5	20	27	20	15	14	9	7	8	20

So you could send the message **13-5-5-20-27-20-15-14-9-7-8-20** to your friend and by using the Numerical Conversion Table, the message could then be decoded. As long as the table used to encode and decode the message is known only to the sender and receiver of the message, it is hoped that unauthorized individuals will be unable to "break" the coded message. The weak link in this method of encryption is that each letter of the alphabet is being replaced by another symbol, preserving the frequency of the individual letters contained in the message. The frequency of each letter in the English alphabet is well known and therefore statistical methods can be used to figure out the message.

For example, the letter *e* occurs most frequently (over 12% of the time) and since most coded messages will be long, an unauthorized individual could begin breaking the code by looking for the number occurring most frequently in the message. This number will most likely correspond to the letter e. Using the known frequencies of the other letters, the rest of the message could then be decoded.

In order to make the secret message more difficult to decode, matrix multiplication is used to further scramble the message before it is sent. The entire process of encoding a message is described and illustrated in the next section.

Encoding a message.

To further scramble our message, use the following steps.

1. Converting characters to integers

The first step in encoding a message is to use the numerical conversion table to convert each letter of the message to a corresponding number (as described earlier).

2. Place the message in a matrix

Pair up the letters as follows:

M E	E T	T	O N	I G	H T
13 5	5 20	27 20	15 14	9 7	8 20

and place the pairs is consecutive columns of matrix $M = \begin{pmatrix} 13 & 5 & 27 & 15 & 9 & 8 \\ 5 & 20 & 20 & 14 & 7 & 20 \end{pmatrix}$.

3. Left-Multiply by an enciphering matrix.

Compute the product of E and M where $E = \begin{pmatrix} 5 & 2 \\ 3 & 1 \end{pmatrix}$ in order to scramble the message more thoroughly. (Matrix E is called an *enciphering matrix.*)

$$S = EM = \begin{pmatrix} 5 & 2 \\ 3 & 1 \end{pmatrix}\begin{pmatrix} 13 & 5 & 27 & 15 & 9 & 8 \\ 5 & 20 & 20 & 14 & 7 & 20 \end{pmatrix} = \begin{pmatrix} 75 & 65 & 175 & 103 & 59 & 80 \\ 44 & 35 & 101 & 59 & 34 & 44 \end{pmatrix}$$

4. Perform modular arithmetic

Now each number in the matrix is replaced with the corresponding letter in the numerical conversion table. To do this, modular 29 arithmetic must be performed on each number in matrix S. It can be shown that

$$\begin{pmatrix} 75 & 65 & 175 & 103 & 59 & 80 \\ 44 & 35 & 101 & 59 & 34 & 44 \end{pmatrix} = \begin{pmatrix} 17 & 7 & 1 & 16 & 1 & 22 \\ 15 & 6 & 14 & 1 & 5 & 15 \end{pmatrix} \pmod{29}$$

5. Determining the encrypted message.

Finally, the matrix $\begin{pmatrix} 17 & 7 & 1 & 16 & 1 & 22 \\ 15 & 6 & 14 & 1 & 5 & 15 \end{pmatrix}$ is placed back into a sequential list and the numerical conversion table is used to determine the letters in the encoded message.

17	15	7	6	1	14	16	1	1	5	22	15
Q	O	G	F	A	N	P	A	A	E	V	O

So the encoded message is **QOGFANPAAEVO**

Exercise 6: Without the help of a computer, encode the message HELP using the enciphering matrix $\begin{pmatrix} 3 & 4 \\ 2 & 3 \end{pmatrix}$.

Encoding a Message with *Mathematica*

Encoding messages by hand is tedious work! A demonstration will now be given of how to use *Mathematica* to encode the following message with the enciphering matrix $\begin{pmatrix} 5 & 3 \\ 7 & 4 \end{pmatrix}$.

MEET AT FIVE P.M. IN THE HOTEL LOBBY.

Begin by assigning a name to the enciphering matrix. Execute the following assignment.

```
encmat = (5 3
          7 4);
```

Next, type in the message to be encoded and assign it the name *msg*. *Be sure to enter only CAPITAL letters and characters found in the numerical conversion table. Also, place the entire message in quotation marks.* The purpose of placing quotation marks around the message is to distinguish the characters from variables and constants. Execute the following assignment.

```
msg = "MEET AT FIVE P.M. IN THE HOTEL LOBBY."
```

The command **MessageToMatrix** will replace each character in *msg* with the corresponding number found in the numerical conversion table and will then place the consecutive pairs of numbers into columns of a matrix. If there are an odd number of characters in the original message, the number 27 (corresponding to the space character) will be placed in the last row and last column of the newly formed matrix. Execute the following command.

```
msgmat = MessageToMatrix[msg]
```

Next, left-multiply by the enciphering matrix and perform modular 29 arithmetic. Execute the following command.

```
codedmat = Mod[encmat . msgmat, 29]
```

The final steps are to place the matrix back into a sequence

{22, 24, 27, 28, 22, 19, 7, 16, 28, 20, 9, 0, 9, 21, 5, 16, 18, 14, 0, 3, 21, 8, 26, 18, 14, 18, 19, 11, 3, 25, 26, 5, 23, 26, 27, 27,15,10} and then replace each number with the corresponding character found in the numerical conversion table. These two steps can be completed using the command **MatrixToMessage** Execute the following statement to assign the coded message the name *secretmsg.*

```
secretmsg = MatrixToMessage[codedmat]
```

So the encoded message to be sent is **VX .VSGP.TI?IUEPRN?CUHZRNRSKCYZEWZ OJ**

Exercise 7: Check your answer to Exercise 6 by using *Mathematica* to encode the message HELP using the enciphering matrix $\begin{pmatrix} 3 & 4 \\ 2 & 3 \end{pmatrix}$.

Exercise 8: Encode the following message using the encyphering matrix $\begin{pmatrix} 9 & 7 \\ 5 & 4 \end{pmatrix}$.

THE HOUSE KEY IS UNDER THE MAT.

The message has already been entered for you, so begin by executing the following assignment statement.

```
newmsg = "THE HOUSE KEY IS UNDER THE MAT."
```

Decoding a Message with Mathematica

Suppose you want to decode the following encrypted message

?NFKWXCWFKDKJ?YBV SA NBQXGDTWGUUFA

which was encoded using the enciphering matrix $\begin{pmatrix} 2 & 3 \\ 5 & 7 \end{pmatrix}$. The process is identical to the process of encoding a message with one exception: after the message has been placed in a matrix using the **MessageToMatrix** command, left-multiply by the inverse of the enciphering matrix (modular 29).

Exercise 9:

(a) Execute the following assignment statement to assign the encoded message the same *secret message* and to assign the modular 29 inverse the name *decmat* (short for deciphering matrix).

(b) Use matrix multiplication and the **Mod** command to verify that *decmat* is the inverse of $\begin{pmatrix} 2 & 3 \\ 5 & 7 \end{pmatrix}$.

(c) Determine the original message.

```
secretmsg = "?NFKWXCWFKDKJ?YBV SA NBQXGDTWGUUFA";
decmat = Inverse[(2 3
                  5 7), Modulus -> 29]
```

Exercise 10: Execute the following assignment statement and then decode the message using the enciphering matrix $\begin{pmatrix} 2 & 3 \\ 5 & 7 \end{pmatrix}$. Follow the instructions of the decoded message using the same enciphering matrix $\begin{pmatrix} 2 & 3 \\ 5 & 7 \end{pmatrix}$.

```
newmsg = "VDTCJJPO IIRN.CWN.HRSQMPB HUFKQDFWIMLO?LUJOAKNTKWXCWH I. NI.V
   BXGWGBFOPKYHDMPVXFWIMLO?L NXCH I.WX ?GWVJ BDYSELFDKFKDKLOLMKYHDHLBQEHTCFWD?
   EHMSFKD?GFAVAUXNBF BKYHDMPDKFKQDFWIMLO?LUJBDDKNJVI NXCAMSQPLH I.
   GFZ PLUKHRSQMPB HUH I.FWIMLO?LUJ"
```

Initialization Cells

Do not evaluate the following commands - they will be automatically evaluated.

```
Off[General::spell1];
```

```
$Post := If[MatrixQ[#], MatrixForm[#], #]&
```

```
Odd::error := "The list entered does not contain an even number of entries."
```

```
ListToMatrix::usage :=
 "ListToMatix[list] places consecutive pairs of numbers in list into
   columns of a matrix. If list contains an odd number of characters,
   the number corresponding to the space character in the numercial
   conversion table is placed in the last column and last row of the matrix."
```

```
ListToMatrix[l_] := Module[{lst},
  lst = l;
  If[OddQ[Length[lst]], lst = Append[lst, 27]];
 Transpose[Partition[lst, 2]]]
```

```
MatrixToList::usage := "MatrixToList[matrix] places consecutive
   pairs of numbers in matrix into a list."
```

```
MatrixToList[m_] :=
 Module[{},
  Flatten[Transpose[m]]]
```

```
CharacterList::usage :=
 "CharacterList is the list of characters in the numerical conversion table."
```

```
CharacterList = {"A", "B", "C", "D", "E", "F", "G", "H", "I", "J", "K", "L", "M",
  "N", "O", "P", "Q", "R", "S", "T", "U", "V", "W", "X", "Y", "Z", " ", ".", "?"};
```

```
AlphaChar::usage := "AlphaChar[i] is the character corresponding to
   the number i in the numerical conversion table."
```

```
AlphaChar[0] = CharacterList[[29]];
```

```
Do[AlphaChar[i] = CharacterList[[i]], {i, 1, Length[CharacterList] - 1}]
```

```
MatchingNumber::usage := "MatchingNumber[char] is the number in the
   numerical conversion table corresponding to character char."
```

```
Do[MatchingNumber[AlphaChar[i]] = i, {i, 0, Length[CharacterList]}]
```

```
NumericalConversion::usage :=
 "NumericalConversion[charlist] converts each character in the list charlist
   into the corresponding number found in the numerical conversion table."
```

```
NumericalConversion[l_] := Map[MatchingNumber, l]
```

```
AlphaConversion::usage :=
 "AlphaConversion[numlist] converts each number in list numlist to
   the corresponding character found in the numerical conversion table."
```

```
AlphaConversion[l_] := Map[AlphaChar, l];
```

```
MessageToMatrix::usage := "MessageToMatrix[msg] replaces each letter in msg
   with the corresponding number found in the numerical conversion table and
   then places each consecutive pairs of numbers in consecutive columns of a
   matrix.  If msg contains an odd number of characters, the number in
   the last row and column of the matrix will be the number 26 which
   corresponds to the space character in the numerical conversion table."
```

```
MessageToMatrix[msg_] := Module[{numlst, nummat, m},
  m = Table[StringTake[msg, {i}], {i, 1, StringLength[msg]}];
  numlst = NumericalConversion[m];
  nummat = ListToMatrix[numlst]]
```

```
MatrixToMessage::usage :=
 "MatrixToMessage mat] takes a 2 by n matrix mat and first forms the list
  {mat[[1,1]],mat[[2,1]],mat[[1,2]],mat[[2,2]],...,mat[[1,n]],
   mat[[2,n]]} and then each number in the list is replaced by the
   correponding character in the Numerical Conversion Table."
```

```
MatrixToMessage[mat_] := Module[{l, msg},
  l = MatrixToList[mat];
  msg = AlphaConversion[l];
  StringJoin[msg]]
```

Computer Assignment: LU Factorization

Purpose: To understand the LU factorization algorithm for square matrices and to solve systems of equations using the LU factorization of a matrix.

Prerequisite Reading: Section 2.5 and page MM-11 in the *Study Guide*.

***Mathematica* Functions:** BackSub, Clear, ForwardSub, Gauss, LUFactor, ReplaceRow, Simplify

Palette Used in this Assignment:

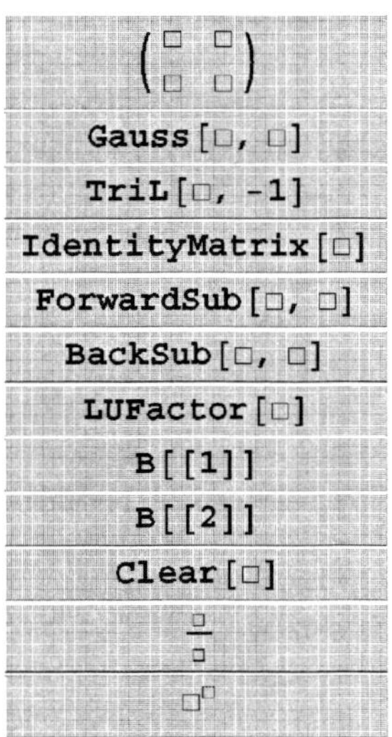

Introduction: The LU factorization of a given $n \times n$ matrix can be quickly found using the **Gauss** command. After reviewing and practicing the LU factorization algorithm, a storage scheme will be examined along with the need for introducing row interchanges in the algorithm.

A Review of the LU Factorization Algorithm

The command **Gauss[A, r]** uses the leading entry in row r of A as a pivot, and uses row replacements to create zeros in the pivot column below this pivot entry. This command will greatly simplify your work in finding the LU factorization of A. For example, suppose you want to find the LU factorization of $\begin{pmatrix} 4 & 10 & 6 & 8 \\ 4 & 8 & 14 & 10 \\ 2 & 4 & 6 & 10 \\ 2 & 4 & 6 & 8 \end{pmatrix}$. Begin by executing the following assignment statement.

$$A = \begin{pmatrix} 4 & 10 & 6 & 8 \\ 4 & 8 & 14 & 10 \\ 2 & 4 & 6 & 10 \\ 2 & 4 & 6 & 8 \end{pmatrix};$$

To create zeros below the number 4 in row 1, execute the following **Gauss** command.

```
Gauss[A, 1]
```

To begin building L, place zeros in all the entries above the main diagonal and place ones on the main diagonal. Next, divide each entry below the pivot position in column 1 by the pivot and place the resulting numbers in the first column of L.

$$L = \begin{pmatrix} 1 & 0 & 0 & 0 \\ 1 & 1 & 0 & 0 \\ \frac{1}{2} & \square & 1 & 0 \\ \frac{1}{2} & \square & \square & 1 \end{pmatrix}.$$

Examine the matrix obtained by computing **Gauss[A,1]** Since the leading entry in row 2 of this matrix is -2, execute the following command to obtain 0's below the -2, after first entering the output number associated with the matrix obtained from the last **Gauss** command.

```
Gauss[%□, 2]
```

Two more entries of L are now known so that $L = \begin{pmatrix} 1 & 0 & 0 & 0 \\ 1 & 1 & 0 & 0 \\ \frac{1}{2} & \frac{1}{2} & 1 & 0 \\ \frac{1}{2} & \frac{1}{2} & \square & 1 \end{pmatrix}$. To obtain the final LU factorization of A, execute the following command after first entering the output number corresponding to the last output.

```
Gauss[% □, 3]
```

Now $L = \begin{pmatrix} 1 & 0 & 0 & 0 \\ 1 & 1 & 0 & 0 \\ \frac{1}{2} & \frac{1}{2} & 1 & 0 \\ \frac{1}{2} & \frac{1}{2} & 1 & 1 \end{pmatrix}$ and $U = \begin{pmatrix} 4 & 10 & 6 & 8 \\ 0 & -2 & 8 & 2 \\ 0 & 0 & -1 & 5 \\ 0 & 0 & 0 & -2 \end{pmatrix}$. To verify that $A = LU$, execute the following cell after entering the output number corresponding to the matrix U obtained in the last output statement. (Note that the semicolon separates the two assignment statements and matrix multiplication is indicated with a period.)

```
L = (1  0  0  0
     1  1  0  0
     1/2 1/2 1 0
     1/2 1/2 1 1); U = % □; L . U
```

Using the LU factorization of A just obtained, an equation such as $\begin{pmatrix} 4 & 10 & 6 & 8 \\ 4 & 8 & 14 & 10 \\ 2 & 4 & 6 & 10 \\ 2 & 4 & 6 & 8 \end{pmatrix} \begin{pmatrix} x_1 \\ x_2 \\ x_3 \\ x_4 \end{pmatrix} = \begin{pmatrix} 0 \\ 0 \\ 4 \\ 2 \end{pmatrix}$ can now be solved using the two step process of solving $L\mathbf{y} = \mathbf{b}$ using the forward substitution command **ForwardSub[L,b]** and then solving $U\mathbf{x} = \mathbf{y}$ by back-substitution using **BackSub[U,y]** Execute the following statement to define **b** and then to find **y** followed by **x**. (Note that the *Mathematica* notation for a vector $\begin{pmatrix} 0 \\ 0 \\ 4 \\ 2 \end{pmatrix}$ is {0, 0, 4, 2}.)

```
b = {0, 0, 4, 2};
y = ForwardSub[L, b];
x = BackSub[U, y]
```

To verify that the solution just obtained is correct, compute the following product.

```
A . x
```

> ⚙Exercise 1: Execute the following assignment statement. Use the **Gauss** command to compute the LU factorization of A and after you use **Gauss** several times to obtain *U*, review your work to find the entries in *L*. Then compute the product of *L* and *U* to verify your work.

```
A = (4  2  3  1
     0  5  0  1
     2  3  6  5
     1  2  3  5);
```

Exercise 2: Use the factorization found in Exercise 1 along with **ForwardSub** and **BackSub** to find a solution to $\begin{pmatrix} 4 & 2 & 3 & 1 \\ 0 & 5 & 0 & 1 \\ 2 & 3 & 6 & 5 \\ 1 & 2 & 3 & 5 \end{pmatrix}\begin{pmatrix} x_1 \\ x_2 \\ x_3 \\ x_4 \end{pmatrix} = \begin{pmatrix} -31 \\ 8 \\ -4 \\ 2 \end{pmatrix}$. Verify your answer by computing the product of A and the solution.

Exercise 3: Execute the following assignment statement. Use the **Gauss** command to compute the LU factorization of A and after you use **Gauss** several times to obtain *U*, review your work to find the entries in *L*. Then compute the product of *L* and *U* to verify your work.

$$A = \begin{pmatrix} 2 & 3 & 0 & 1 \\ 4 & 5 & 2 & 2 \\ 6 & 9 & -3 & 4 \\ 0 & 2 & -1 & 5 \end{pmatrix};$$

Exercise 4: Use the LU factorization obtained in Exercise 3 to solve the matrix equation

$$\begin{pmatrix} 2 & 3 & 0 & 1 \\ 4 & 5 & 2 & 2 \\ 6 & 9 & -3 & 4 \\ 0 & 2 & -1 & 5 \end{pmatrix}\begin{pmatrix} x_1 \\ x_2 \\ x_3 \\ x_4 \end{pmatrix} = \begin{pmatrix} 2 \\ 1 \\ 7 \\ 5 \end{pmatrix}.$$

Storage of LU Factorization

Suppose you wish to store the LU factorization of a given matrix A for later use. One way to do this would be to simply store matrices L and U separately. But this would not be efficient use of storage spaces since so many of the entries in L and U are zeros or ones. A more efficient storage scheme would be to replace the zeros below the main diagonal of U with the entries below the main diagonal of L

For instance, an LU factorization of $A=\begin{pmatrix}4 & 10 & 6 & 8\\ 4 & 8 & 14 & 10\\ 2 & 4 & 6 & 10\\ 2 & 4 & 6 & 8\end{pmatrix}$ is $L=\begin{pmatrix}1 & 0 & 0 & 0\\ 1 & 1 & 0 & 0\\ \frac{1}{2} & \frac{1}{2} & 1 & 0\\ \frac{1}{2} & \frac{1}{2} & 1 & 1\end{pmatrix}$ and $U=\begin{pmatrix}4 & 10 & 6 & 8\\ 0 & -2 & 8 & 2\\ 0 & 0 & -1 & 5\\ 0 & 0 & 0 & -2\end{pmatrix}$. For storage purposes, let $U=\begin{pmatrix}4 & 10 & 6 & 8\\ 1 & -2 & 8 & 2\\ \frac{1}{2} & \frac{1}{2} & -1 & 5\\ \frac{1}{2} & \frac{1}{2} & 1 & -2\end{pmatrix}$.

To illustrate how to store the LU factorization with *Mathematica*, without retyping in all the entries, begin by executing the following assignments statements.

$$L=\begin{pmatrix}1 & 0 & 0 & 0\\ 1 & 1 & 0 & 0\\ \frac{1}{2} & \frac{1}{2} & 1 & 0\\ \frac{1}{2} & \frac{1}{2} & 1 & 1\end{pmatrix};\ U=\begin{pmatrix}4 & 10 & 6 & 8\\ 0 & -2 & 8 & 2\\ 0 & 0 & -1 & 5\\ 0 & 0 & 0 & -2\end{pmatrix};$$

The command **TriL[A,-1]** creates a matrix whose entries below the main diagonal are the entries of A and whose remaining entries are all zero. (If you are interested in a more general definition of **TriL[A,k]**, then execute the command **?TriL**.) Execute the following assignment statement to create the storage matrix.

```
U = TriL[L, -1] + U
```

To reconstruct matrices L and U from the storage matrix, execute the following command. (**IdentityMatrix[4]** is the 4×4 identity matrix which is needed to place ones on the diagonal of L.)

```
L = TriL[U, -1] + IdentityMatrix[4]
```

Convince yourself that the following assignment statement will form U such that $A=$ LU. Then execute the assignment statement.

```
U = U - TriL[U, -1]
```

Exercise 5: Form a storage matrix for the LU factorization of matrix A found in Exercise 1.

Exercise 6: Suppose the storage matrix for an LU factorization of matrix A is given in the following input cell. Execute the assignment statement and then determine matrix A by first finding L and U such that $A = LU$.

$$U = \begin{pmatrix} 22 & 0 & 25 & -80 \\ 32 & 52 & -49 & -3 \\ 1 & 30 & 6 & 0 \\ 92 & 80 & 56 & 40 \end{pmatrix}$$

The LUFactor command

A *Mathematica* function for determining the LU factorization of a matrix is **LUFactor.** For example, to determine the LU factorization of $\begin{pmatrix} 2 & 1 \\ 1 & 3 \end{pmatrix}$, execute the following command.

$$B = \text{LUFactor}\left[\begin{pmatrix} 2 & 1 \\ 1 & 3 \end{pmatrix}\right]$$

The value of **B[[1]]** will be the storage matrix and **B[[2]]** is a pivot row list to be discussed later on in this assignment. Execute the following command to see the storage matrix.

```
B[[1]]
```

From the storage matrix, it follows that $L = \begin{pmatrix} 1 & 0 \\ \frac{1}{2} & 1 \end{pmatrix}$ and $U = \begin{pmatrix} 2 & 1 \\ 0 & \frac{5}{2} \end{pmatrix}$. Execute the following commands to obtain the factorization of L and U and to verify that $LU = \begin{pmatrix} 2 & 1 \\ 1 & 3 \end{pmatrix}$.

```
L = TriL[B[[1]], -1] + IdentityMatrix[2];
U = B[[1]] - TriL[B[[1]], -1];
L . U
```

Exercise 7: Use **LUFactor** to find L and U for matrix $\begin{pmatrix} 4 & 1 & 3 \\ 2 & 4 & 7 \\ 1 & 2 & 3 \end{pmatrix}$. Compute $L\,.\,U$ to verify that the product of L and U does indeed equal the original matrix.

A General Formula for the 3 × 3 Case

Assume $A = \begin{pmatrix} a & b & c \\ d & e & f \\ g & h & i \end{pmatrix}$ can be row reduced to the identity matrix without any row interchanges. Then the **LUFactor** command can be used to find a general formula for the storage matrix . Execute the following commands.

```
Clear[a, b, c, d, e, f, g, h, i]; B = LUFactor[(a b c
                                                 d e f
                                                 g h i)]
```

Execute the following command to see the storage matrix:

```
B[[1]]
```

Now use the **Simplify** command to eliminate the compound fractions appearing in the storage matrix. Execute the following statement.

```
B[[1]] = Simplify[B[[1]]]
```

Exercise 8: Use the storage matrix just found to obtain L and U such that $A = LU$.

Exercise 9: If $a = 0$ in matrix A, then the storage matrix formula does not work. Why? What assumption is being made with implies $a \neq 0$? Why?

Type your answers here:

Exercise 10: Further examination of the storage matrix shows that we must also assume that $-bd + ae \neq 0$. Under the current assumption of no row interchanges, explain why this quantity cannot be 0 *Hint: Recall that the command **ReplaceRow[A,m,c,n]** will replace row m by its sum with c times row n. Execute the following input statement and then use the result of the row operation to explain your reasoning.*

Type your answers here:

```
ReplaceRow[{{a, b, c}, {d, e, f}, {g, h, i}}, 2, -d/a, 1]
```

$$\text{ReplaceRow}\left[\begin{pmatrix} a & b & c \\ d & e & f \\ g & h & i \end{pmatrix}, 2, \frac{-d}{a}, 1\right]$$

Permutation Matrices

From the previous two exercises, it is clear that the LU factorization algorithm, as discussed so far, is insufficient for the case where row interchanges are necessary. Even in the case where row interchanges are unnecessary, serious roundoff errors can occur in the process of obtaining L and U if a pivot is relatively small when compared to the other numbers in the pivot column. Therefore the algorithm needs to be modified to allow for row interchanges. Before discussing this modification, a quick review of permutation matrices is in order. Suppose the $n \times n$ matrix P is obtained from the $n \times n$ identity matrix I by a series of row interchanges. Then for any $n \times n$ matrix A, suppose the same row switches that were applied to I to obtain P are applied to A to obtain B. Then it turns out that $B = PA$. Matrix P is an example of a *permutation matrix.*

As a demonstration, observe the contents of the matrix P in the following cell. Then note that row 1 of the identity matrix is in row 3 of P, row 2 of I is in row 1 of P and row 3 of I is in row 2 of P. Execute the following command and study the contents of the product PA to see that the same row switches performed on I to obtain P are performed on A to obtain the product PA.

```
Clear[a, b, c, d, e, f, g, h, i];
```

$$A = \begin{pmatrix} a & b & c \\ d & e & f \\ g & h & i \end{pmatrix};\ P = \begin{pmatrix} 0 & 1 & 0 \\ 0 & 0 & 1 \\ 1 & 0 & 0 \end{pmatrix};\ P.A$$

Exercise 11: Determine a permutation matrix P such that $PA=\begin{pmatrix} g & h & i \\ a & b & c \\ d & e & f \end{pmatrix}$. Verify your work with *Mathematica.*

LU Factorization with Partial Pivoting

The general LU factorization algorithm usually includes a strategy, called *partial pivoting,* where rows are interchanged so that a pivot having the largest possible absolute value is obtained (see pages 19 and 20 in your textbook). The algorithm for partial pivoting is introduced in the *Study Guide.* Here the **LUFactor** function is used to find the LU factorization of a given square matrix A with partial pivoting. Using this *Mathematica* function, you can obtain the permutation matrix P along with L and U such that $PA = LU$.

To see how to find L and U for a particular example, execute the following cell to find an LU factorization related to

$$\begin{pmatrix} 2 & 2 & 4 \\ 4 & 2 & 0 \\ 6 & 2 & 2 \end{pmatrix}.$$

```
     (2  2  4)
A =  (4  2  0) ; B = LUFactor[A]
     (6  2  2)
```

The last list in the result, {3, 1, 2} indicates the permutation matrix P is obtained by placing row 3 of the identity matrix in row 1 of P, row 1 of the identity matrix in row 2 of P and row 2 of I in row 3 of P. In the following cell, $\begin{pmatrix} 0 & 0 & 1 \\ 1 & 0 & 0 \\ 0 & 1 & 0 \end{pmatrix}$ is assigned to P and then the product of P and **B[[1]]** is computed to obtain the storage matrix for L and U.

```
     (0  0  1)
P =  (1  0  0) ; P.B[[1]]
     (0  1  0)
```

From the storage matrix, you now know L and U. The following statement assigns matrices to L and U and then displays the product LU. Execute the following input cell after entering the number corresponding to the last output statement.

```
Clear[L, U];
L = TriL[%□, -1] + IdentityMatrix[3] ;
U = %□ - TriL[%□, -1];
L . U
```

Execute the following command to verify that P.A=L.U.

```
P = {{0, 0, 1}, {1, 0, 0}, {0, 1, 0}}; P.A
```

$$P=\begin{pmatrix}0&0&1\\1&0&0\\0&1&0\end{pmatrix};\ P.A$$

Suppose you wish to solve A**x**=**b** where **b**= $\begin{pmatrix}6\\-2\\-4\end{pmatrix}$ using the LU factorization for PA. Since PA**x**=P**b** and PA=LU, then LU**x**= P**b.** Therefore to find a solution **x**, first solve L**y**=P**b** for **y** and then solve U**x**=**y** for **x**. Execute the following cell.

```
b = {6, -2, -4};
y = ForwardSub[L, P.b];
x = BackSub[U, y]
```

To verify the solution, execute the following product to make sure the value of **b** is obtained.

```
A.x
```

Exercise 12: Find matrices P, L and U such that PA=LU where L and U provide the LU factorization of $A=\begin{pmatrix}4&4&0&-4\\2&0&0&0\\-4&1&0&4\\4&4&2&0\end{pmatrix}$. Verify that PA=LU.

Exercise 13. Use the LU factorization found in Exercise 12 to solve the matrix equation

$$\begin{pmatrix}4&4&0&-4\\2&0&0&0\\-4&1&0&4\\4&4&2&0\end{pmatrix}\begin{pmatrix}x_1\\x_2\\x_3\\x_4\end{pmatrix}=\begin{pmatrix}4\\10\\-19\\12\end{pmatrix}$$

Initialization Cell

```
Off[General::spell1];
Needs["LinearAlgebra`GaussianElimination`"];
Needs["LayData`LayFunctions`"];
$Post := If[MatrixQ[#] || VectorQ[#], MatrixForm[#], #]&;

TriL::usage = "TriL[A] produces the lower triangular part of any matrix A with
    zeros above the main diagonal.  TriL[A,k] selects all the elements on and
    below the kth diagonal of A , with k<0 below the main diagonal
    and k>0 above the main diagonal.  Also, TriL[A,0]=TriL[A].";

TriL[A_, k_: 0] := Module[{m, n},
  {m, n} = Dimensions[A];
  Table[If[i + k < j, 0, A[[i, j]]], {i, m}, {j, n}]];

ForwardSub[a_, b_] := Module[{n, A, B},
  n = Length[b]; B = b; A = a;
  For[k = 1, k ≤ n - 1, k++,
   For[i = k + 1, i ≤ n, i++,
    m = A[[i, k]] / A[[k, k]];
    For[j = k + 1, j ≤ n, j++,
     A[[i, j]] = A[[i, j]] - m A[[k, j]]];
    B[[i]] = B[[i]] - m B[[k]]]];
  Return[B]];

BackSub[a_, b_] := Module[{n, A, B, x},
  n = Length[b]; A = a; B = b;
  x = Table[0, {n}];
  For[i = n, i ≥ 1, i--,
   x[[i]] = B[[i]];
   For[j = i + 1, j ≤ n, j++,
    x[[i]] = x[[i]] - A[[i, j]] x[[j]]];
   x[[i]] = x[[i]] / A[[i, i]]];
  Return[x]];
```

Computer Assignment: Computer Graphics

Purpose: To create images using homogeneous coordinates and to use matrix multiplication to stretch, rotate and translate images.

Prerequisite Reading: Section 2.8

***Mathematica* Functions:** Balloon, Clear, Cos, HGMatrix, PlotImage, Sin

Palette Used in this Assignment:

$\begin{pmatrix} 1 & 0 & \square \\ 0 & 1 & \square \\ 0 & 0 & 1 \end{pmatrix}$
$\begin{pmatrix} \text{Cos}[\square] & -\text{Sin}[\square] & 0 \\ \text{Sin}[\square] & \text{Cos}[\square] & 0 \\ 0 & 0 & 1 \end{pmatrix}$
$\begin{pmatrix} \square & 0 & 0 \\ 0 & \square & 0 \\ 0 & 0 & 1 \end{pmatrix}$
$\begin{pmatrix} \square & \square & \square \\ \square & \square & \square \\ \square & \square & \square \end{pmatrix}$
`HGMatrix[□]`
`PlotImage[□, □, □, □, □]`
`Cos[□]`
`Sin[□]`
`Clear[□]`
θ
$\frac{\square}{\square}$
π
α
β

Introduction: As stated in the textbook, computer graphics is closely related to matrix multiplication. But translating (i.e. moving) an object from one point to another on the screen does not directly correspond to a matrix multiplication. To avoid this problem, *homogeneous coordinates* will be used where each ordered pair (x, y) is associated with the ordered triple $(x, y, 1)$. Translations can then be performed using multiplication by 3×3 matrices.

Creating an Image

Suppose you want to create the image of the following figure with *Mathematica*.

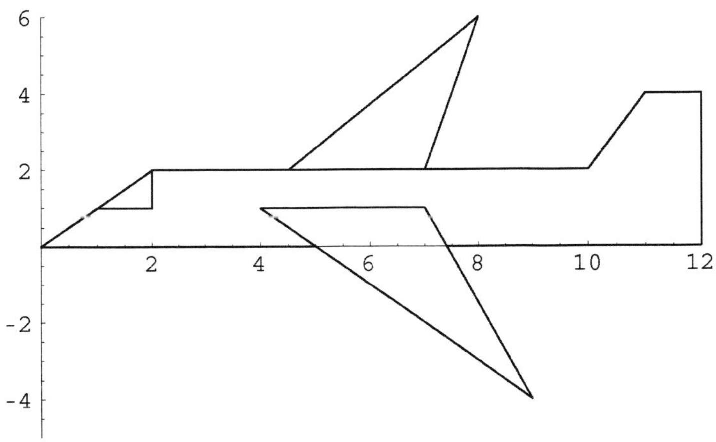

To create this image, the points appearing in the next figure must be connected together in the following order:

(0,0), (5,0), (9,-4), (7,1), (4,1), (5,0), (4,1), (7,1), (7.4,0), (12,0), (12,4), (11,4), (10,2), (7,2), (8,6), (4.5,2), (7,2), (2,2), (2,1), (1,1), (2,2), (0,0)

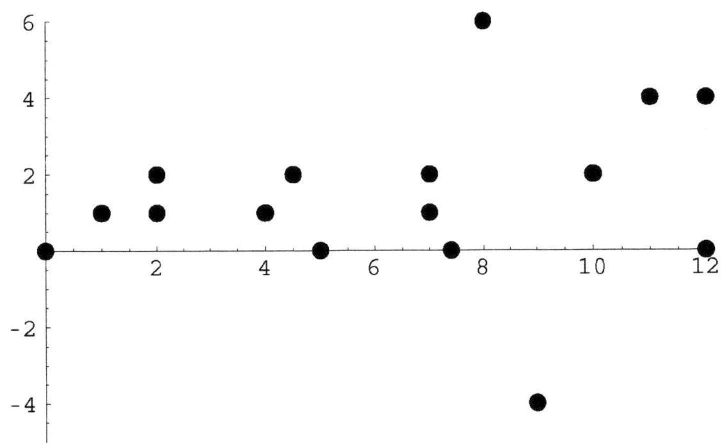

The *Mathematica* notation for an ordered pair (*x*, *y*) is {*x*, *y*}. The following assignment statement creates a list containing the points appearing in the jet image. *The points must appear in the order in which they will be connected together.* The list is assigned the name *jet.*

The command **HGMatrix[jet]** creates a matrix whose nth column is $\begin{pmatrix} x \\ y \\ 1 \end{pmatrix}$, where the point {*x*, *y*} is the nth ordered pair appearing in *jet*. This matrix is assigned the name *jetmat*. Execute the following input cell.

```
jet = {{0, 0}, {5, 0}, {9, -4}, {7, 1}, {4, 1}, {5, 0},
  {4, 1}, {7, 1}, {7.4, 0}, {12, 0}, {12, 4}, {11, 4}, {10, 2}, {7, 2},
  {8, 6}, {4.5, 2 }, {7, 2}, {2, 2}, {2, 1}, {1, 1}, {2, 2}, {0, 0}};

jetmat = HGMatrix[jet]
```

The command **PlotImage[jetmat, a, b, c, d]** is used to plot and connect the ordered pairs (*x*, *y*) appearing in jetmat from $x = a$ to $x = b$ and from $y = c$ to $y = d$. Execute the following statement.

```
PlotImage[jetmat, 0, 17, -10, 6]
```

Exercise 1: Complete the following steps to plot the image of the letter N found in Figure 1 on page 155 in your textbook.

(a) The points used to create the letter N in Figure 1 are listed in column form in matrix D on the bottom of page 155. Create a list, called *letN,* containing a list of points needed to create the letter. Be sure to list the points in the order in which they will be connected together. The created list should be begin and end with {0,0} in order to connect the last point back to the first point.

(b) Use the command **HGMatrix** to convert *letN* into a matrix and assign this matrix the name *letNmat* and then use the **PlotImage** command to plot the corresponding image.

Translating Images

Read Example 4 appearing on page 157 in your textbook. Left multiplying *jetmat* by $\begin{pmatrix} 1 & 0 & 5 \\ 0 & 1 & -3 \\ 0 & 0 & 1 \end{pmatrix}$ will move the entire image 5 units right and 3 units down.

For example, to move the previously created jet 5 units to the right and 3 units down, first execute the following product (recall that a period is used to indicate matrix multiplication).

```
newjetmat = (1  0   5)
            (0  1  -3) . jetmat
            (0  0   1)
```

Now perform the following **PlotImage** command to graph the translated image.

```
PlotImage[newjetmat, 0, 17, -10, 6]
```

To instead multiply and plot all in one step, execute the following command.

```
PlotImage[(1  0  5
           0  1 -3
           0  0  1).jetmat, 0, 17, -10, 6]
```

Exercise 2: Show the image of *jetmat* after it has been shifted 2 units right and 4 units down. Perform the multiplication and the plot all in one step

Rotating Images

Read Example 5 on page 157. Left multiplying *jetmat* by $\begin{pmatrix} \text{Cos}[\theta] & -\text{Sin}[\theta] & 0 \\ \text{Sin}[\theta] & \text{Cos}[\theta] & 0 \\ 0 & 0 & 1 \end{pmatrix}$, where θ is in radians, will rotate the image counterclockwise about the origin by an angle of θ. For example, execute the following command to rotate the jet 30 degrees (i.e., $\frac{\pi}{6}$ radians) counterclockwise about the origin.

```
PlotImage[(Cos[-π/6]  -Sin[-π/6]  0
           Sin[-π/6]   Cos[-π/6]  0
               0           0      1).jetmat, 0, 17, -10, 6]
```

Scaling Images

Exercise 3: Show the image of the letter N your created in the first exercise after first scaling the x-coordinates by $\frac{1}{2}$ and the y-coordinates by 2. *(Hint: See Example 5 on page 157 in your textbook.)*

Reflecting Images

Exercise 4: Plot the image of the product $\begin{pmatrix} -1 & 0 & 0 \\ 0 & 1 & 0 \\ 0 & 0 & 1 \end{pmatrix}$. *jetmat* to obtain the reflection of the jet about the y axis. Type in an explanation below of why the matrix $\begin{pmatrix} -1 & 0 & 0 \\ 0 & 1 & 0 \\ 0 & 0 & 1 \end{pmatrix}$ produces this particular reflection.

Explanation:

Exercise 5: What matrix would produce a reflection about the x-axis? (Fill in the blank boxes below.) Plot the reflection of the letter N you created earlier about the x-axis to verify your work.

Matrix: $\begin{pmatrix} \square & \square & \square \\ \square & \square & \square \\ \square & \square & \square \end{pmatrix}$

Combining Translations, Reflections, and Rotations

Before continuing, read the Practice Problem starting on the bottom of page 162 in your textbook. Be sure you understand the solution to this practice exercise before continuing.

Execute the following command to see the image represented by the matrix called **Balloon**.

```
PlotImage[Balloon, -60, 60, -40, 40]
```

Notice that the balloon is on its side. Suppose you wish to place the following balloon in an upright position. One way to do this would be to rotate the image 90 degrees clockwise about the center of the balloon (the center of the balloon is (7, 21)) to obtain the following image.

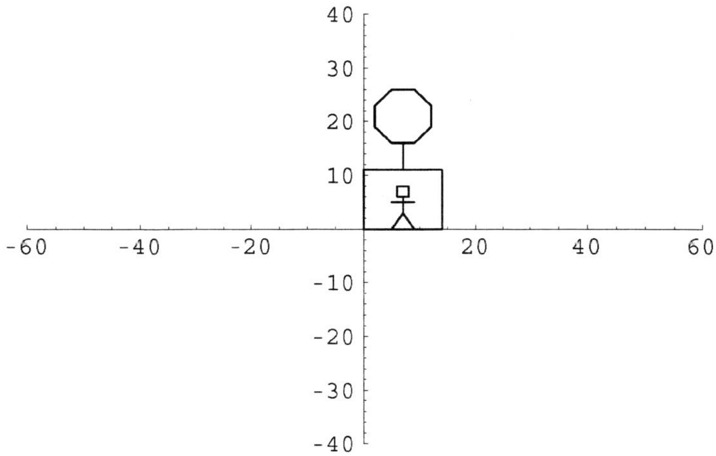

This rotation can be accomplished by first translating the image in Figure 1 by (−7, −21) (see Figure 2), then rotating 90 degrees clockwise about the origin (see Figure 3) and then translating back by (7, 21) (see Figure 4).

Figure 1:

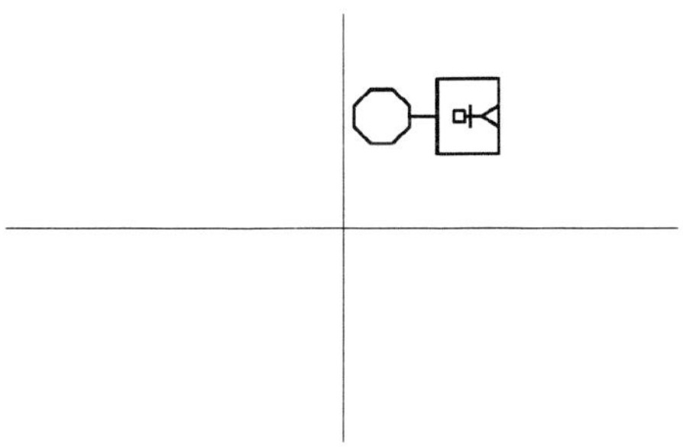

Figure 2:

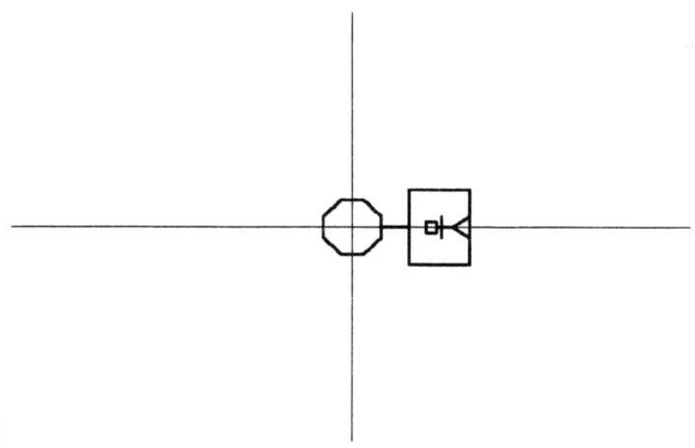

Figure 3:

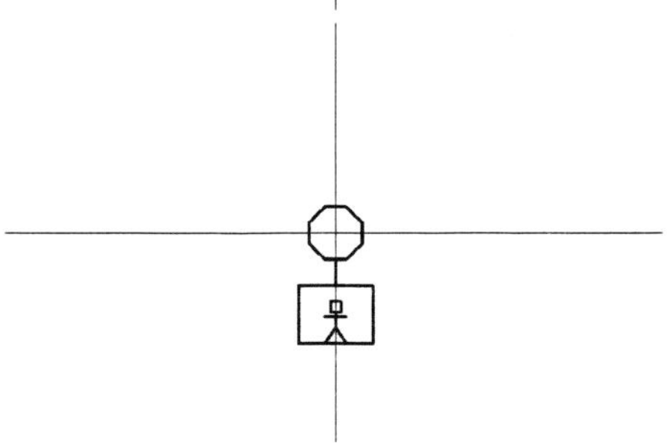

Figure 4:

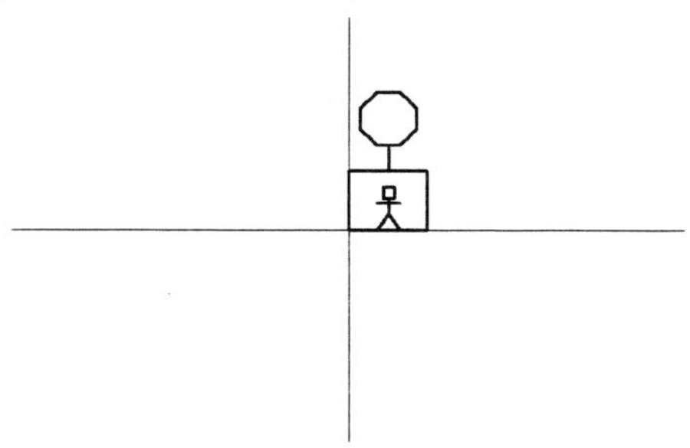

Exercise 6: Construct a 3×3 matrix called *matR* that can be used to rotate the balloon in Figure 1 90 degrees clockwise about the point (7, 21).

Exercise 7: To verify your work to Exercise 6, plot the image of the product *matR*.*Balloon* using the **PlotImage** command.

⚙Exercise 8: Construct a 3×3 matrix called *rev* that can be used to turn around the jet as shown below.

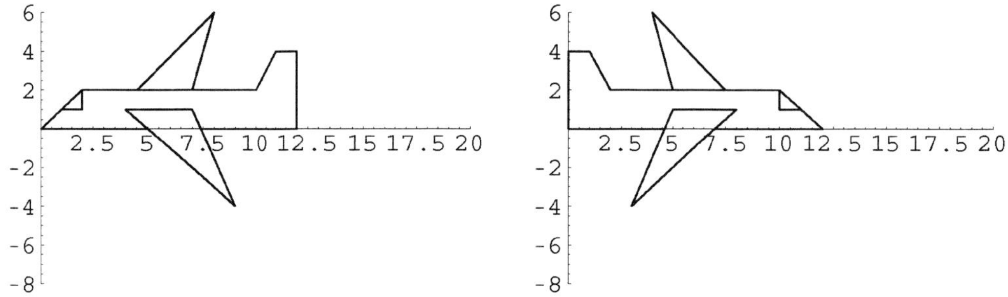

⚙Exercise 9: To verify your work to Exercise 8, assign the name *revjetmat* to the product of *rev . jetmat* and then use the **PlotImage** command to plot the corresponding image represented by *revjetmat*.

Suppose you to first wish to rotate the image represented by *revjetmat* counterclockwise about the origin though an angle of θ and then move the object a distance of r units at an angle of θ (see the following figures). After the rotation, the image would need to be translated by $(r\cos\theta,\ r\sin\theta)$.

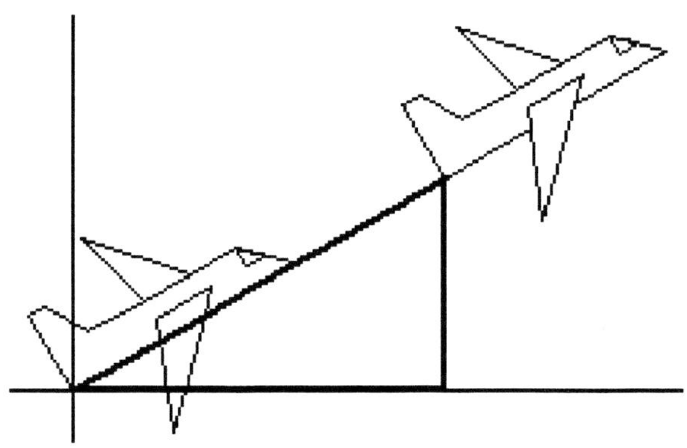

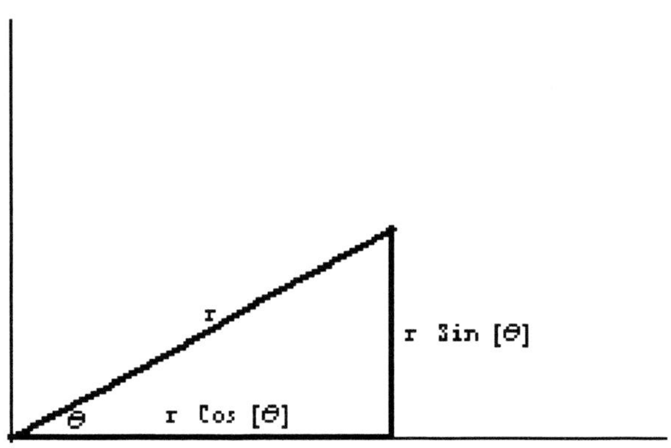

Exercise 10: Find a 3×3 matrix called *move* which will rotate an object about the origin at an angle of 30 degrees and will then move the image 10 units at an angle of 30 degrees. Verify your work using *revjetmat*.

Exercise 11: First create a matrix called *newjetmat* by executing the assignment statement in the cell immediately following this exercise cell. Create a matrix called *drop* that will rotate an image about the point (5,12) clockwise through an angle of 60 degrees and will then move the object 20 units at this downward angle (see the following figure). Then verify your work by left multiplying *newjetmat* by *drop* and plotting the corresponding image.

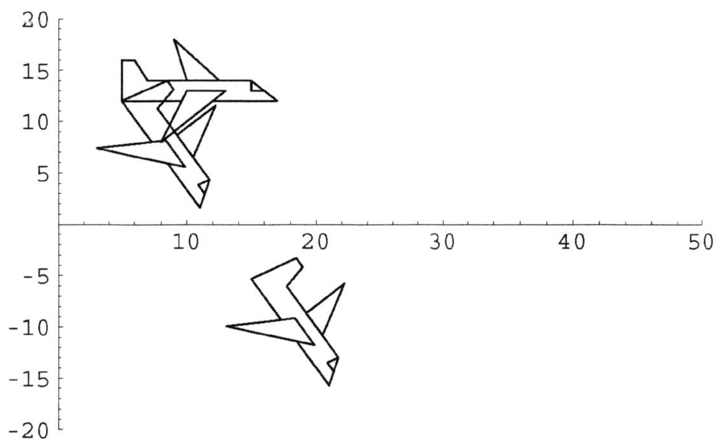

$$\text{newjetmat} = \begin{pmatrix} 1 & 0 & 5 \\ 0 & 1 & 12 \\ 0 & 0 & 1 \end{pmatrix}.\text{jetmat}$$

General Formulas

You can take advantage of the symbolic power of *Mathematica* to come up with general formulas for some the composite transformations you worked with earlier in this assignment. For example, to obtain a matrix used to translate an object by $(-a, -b)$, then to rotate about the origin by an angle of θ and finally translating back by (a, b), execute the following commands.

```
Clear[a, b, θ];
```

$$\begin{pmatrix} 1 & 0 & a \\ 0 & 1 & b \\ 0 & 0 & 1 \end{pmatrix} . \begin{pmatrix} Cos[\theta] & -Sin[\theta] & 0 \\ Sin[\theta] & Cos[\theta] & 0 \\ 0 & 0 & 1 \end{pmatrix} . \begin{pmatrix} 1 & 0 & -a \\ 0 & 1 & -b \\ 0 & 0 & 1 \end{pmatrix}$$

To verify your work to Exercise 6, assign the following values to a, b and θ. Recall that **%n** refers to the contents of output statement **Out[n]**. So replace the empty box in %□ with the output number associated with the matrix you just found. Compare the result with Exercise 6.

$$a = 7;\ b = 21;\ \theta = -\frac{\pi}{2};$$

```
%□
```

> ❀Exercise 12: Find a matrix that will rotate an figure about the origin at an angle of θ and will then move the object a distance or r units at an angle of θ. Be sure to use the **Clear** command first to clear out any variables that will appear in your matrices. Then assign values to the unknown variables to verify your work to Exercise 10. (The figure illustrating this problem was given in Exercise 10.)

Recall that $\begin{pmatrix} Cos[\theta] & -Sin[\theta] & 0 \\ Sin[\theta] & Cos[\theta] & 0 \\ 0 & 0 & 1 \end{pmatrix}$ is used to rotate an object about the origin at an angle of θ. Therefore to rotate an object about the origin at an angle of $\alpha+\beta$, you would use the matrix $\begin{pmatrix} Cos[\alpha+\beta] & -Sin[\alpha+\beta] & 0 \\ Sin[\alpha+\beta] & Cos[\alpha+\beta] & 0 \\ 0 & 0 & 1 \end{pmatrix}$. By forming this matrix in another way, we can find formulas for Cos[α+β] and Sin[α+β].

> ❀Exercise 13: Use matrix multiplication to find a matrix which will first rotate an object about the origin through an angle of α and will then continue to rotate the object another β radians about the origin. Compare the resulting matrix with $\begin{pmatrix} Cos[\alpha+\beta] & -Sin[\alpha+\beta] & 0 \\ Sin[\alpha+\beta] & Cos[\alpha+\beta] & 0 \\ 0 & 0 & 1 \end{pmatrix}$ to find formulas for Cos[α+β] and Sin[α+β]. Fill in the missing boxes below:
>
> Cos[α+β]=□
>
> Sin[α+β]=□

Initialization Files

The following commands will be executed automatically - do not reexecute the following commands.

```
Off[General::spell1];
```

```
$Post := If[MatrixQ[#] || VectorQ[#], MatrixForm[#], #]&
```

```
Needs["LinearAlgebra`MatrixManipulation`"];
```

```
HGMatrix[mat_] := Transpose[AppendRows[mat, Table[{1}, {Length[mat]}]]];
```

```
PlotImage[mat_, xmin_, xmax_, ymin_, ymax_] :=
  ListPlot[Transpose[TakeRows[mat, 2]], PlotJoined -> True, Axes -> Automatic,
   AxesOrigin -> {0, 0}, PlotRange -> {{xmin, xmax}, {ymin, ymax}}];
```

$$\text{Balloon} = \begin{pmatrix} 1 & 0 & 7 \\ 0 & 1 & 21 \\ 0 & 0 & 1 \end{pmatrix} . \begin{pmatrix} \text{Cos}[\pi/2] & -\text{Sin}[\pi/2] & 0 \\ \text{Sin}[\pi/2] & \text{Cos}[\pi/2] & 0 \\ 0 & 0 & 1 \end{pmatrix} . \begin{pmatrix} 1 & 0 & -7 \\ 0 & 1 & -21 \\ 0 & 0 & 1 \end{pmatrix} .\text{Transpose}[$$

```
  {{0, 0, 1}, {5, 0, 1}, {7, 3, 1}, {7, 5, 1}, {5, 5, 1}, {9, 5, 1}, {7, 5, 1},
   {7, 6, 1}, {6, 6, 1}, {6, 8, 1}, {8, 8, 1}, {8, 6, 1}, {7, 6, 1}, {7, 3, 1},
   {9, 0, 1}, {4, 0, 1}, {14, 0, 1}, {14, 11, 1}, {0, 11, 1}, {0, 0, 1},
   {0, 11, 1}, {7, 11, 1}, {7, 16, 1}, {9, 16, 1}, {12, 19, 1}, {12, 23, 1},
   {9, 26, 1}, {5, 26, 1}, {2, 23, 1}, {2, 19, 1}, {5, 16, 1}, {7, 16, 1}}];
```

Computer Assignment: Motion Pictures

Purpose: To use the transformations introduced in the computer graphics assignment to create simple motion pictures.

Prerequisite Reading: Section 2.8, Computer Graphics Assignment

***Mathematica* Functions:** : Clear, Cos, PlotImage, Rot, Scale, Sin, Table

```
(1 0 □)
(0 1 □)
(0 0 1)
Rot[□, □, □]
Scale[□, □, □, □]
frame[i_] := (frame[i] = □)
frame[i_] :=
  (frame[i] = □) /; □ ≤ i ≤ □
frame[0] =
frame[i]
Table[□, {□, □, □}]
PlotImage[□, □, □, □, □]
Balloon
jetmat
Clear[□]
Cos[□]
Sin[□]
π
□/□
```

Introduction: In the previous computer assignment, matrix multiplication was used to produce rotations, translations, reflections and scalings of images. These transformations can be applied repeatedly to produce frames of a simple motion picture and then these frames can be animated using *Mathematica*.

Warning!! *Computer graphics images require large amounts of memory. In order to save memory space, it is highly recommended that you delete the images in this assignment after viewing your work.*

Creating and Viewing a Motion Picture Frames

In this assignment, **Balloon** is a matrix containing homogeneous coordinates for the image of a balloon. Execute the following **PlotImage** command to see the image. (The semicolon on the end of the statement suppresses the output label *Graphics* from being printed.)

```
PlotImage[Balloon, 0, 100, 0, 60];
```

Recall from the previous assignment that $\begin{pmatrix} 1 & 0 & s \\ 0 & 1 & t \\ 0 & 0 & 1 \end{pmatrix}$ is used to translate an image s units horizontally and t units vertically. Suppose you want to create a motion picture of the balloon moving upward along a line making a 45 degree angle with the ground. To do this, you can create a sequence of movie "frames" that you view in rapid succession, with the balloon translated a small amount from frame to frame. To accomplish this task, execute the input cell below to:

1. Clear any previous definitions of **frame.**

2. Define **frame[0]** to equal the matrix **Balloon.**

3. Set up a recursive equation which defines **frame[** *i* **]** in terms of the previous frame. At each step, the image in **frame[i]** is obtained by translating the image in **frame[i-1]**, one unit horizontally and one unit vertically. (The underscore lets *Mathematica* know that *i* is a variable and := is a delayed assignment statement that must be used when a function is defined recursively.)

4. Display the results of **PlotImage[frame[*i*], 0, 100, 0, 60]** for $i = 0, 1, 2, \ldots, 20$ using the **Table** command. (The semicolon suppresses the graphics output labels from being displayed.)

```
Clear[frame];
frame[0] := Balloon;
             (1  0  1)
frame[i_] := (0  1  1) . frame[i - 1];
             (0  0  1)
Table[PlotImage[frame[i], 0, 100, 0, 60], {i, 0, 20}];
```

A sequence of 21 frames has now been produced. For each frame, there is a cell bracket in the right margin representing that particular frame. Immediately to the right of the 21 frame brackets is a long bracket representing the 21 frames. Double-click this long frame bracket now to "stack" the frames on top of each other so that only the first frame is visible. The second bracket to the right should be highlighted (if you accidentally selected the wrong bracket, double click on the selected bracket to undo your work and try again). From the menu, select **Cell** followed by **Animate Selected Graphics** to view the motion picture. While the cells are being animated, an animation toolbar is displayed in the bottom, left corner of the notebook. The two right buttons in the toolbar will allow you to speed up and slow down the animation. Left click on the slow down button a few times to slow down the animation speed. In Exercise 1, determine the effect of each animation button by pressing it while the animation is taking place.

Exercise 1: Describe the purpose of the other buttons in the animation toolbar through trial and error.

Enter description here:

For the purpose of saving space, delete the picture frames used to create the last movie, but **do not** delete the commands used to create the frames (just in case you want to reexecute these commands again later). An easy way to delete the frames is to select the bracket associated with the 21 frames and then press **Delete**.

Building Frames more Efficiently

The command **frame[*i*_] :=** $\begin{pmatrix} 1 & 0 & 1 \\ 0 & 1 & 1 \\ 0 & 0 & 1 \end{pmatrix}$ **. frame[*i* – 1]** (found in the last input cell) is an inefficient way of defining a function recursively. To see why, you need to understand what *Mathematica* is doing when it executes **frame[1], frame[2], frame[3]** and so forth. When **frame[1]** is executed, the function **frame[0]** is executed and then **frame[1]** is created using the product $\begin{pmatrix} 1 & 0 & 1 \\ 0 & 1 & 1 \\ 0 & 0 & 1 \end{pmatrix}$ **. frame[0]**. When **frame[2]** is executed, the commands **frame[0]** and then **frame[1]** are executed first, even if their values have already been determined, and then **frame[2]** is created using the product $\begin{pmatrix} 1 & 0 & 1 \\ 0 & 1 & 1 \\ 0 & 0 & 1 \end{pmatrix}$ **. frame[1]**. In general, when **frame[*n*]** is executed, the commands **frame[0], frame[1], ..., frame[*n* – 1]** are all executed first, even if their have already been created! Since you are creating a table of values, when **frame[*n*]** is executed, the value of **frame[*n* – 1]** will already be known, so in order to keep *Mathematica* from reexecuting **frame[0], frame[1], ..., frame[n-1],** the command

frame[*i*_] := $\left(\textbf{frame}[i] = \begin{pmatrix} 1 & 0 & 1 \\ 0 & 1 & 1 \\ 0 & 0 & 1 \end{pmatrix} . \textbf{frame}[i-1] \right)$

is used instead in define **frame[i]**. Compare the following input cell with the previous input cell and then execute the following input cell. (If you are using a slower computer, you may notice that it takes less time to create the frames with this input cell!) After the frames have been created, animate your movie and then delete the picture frames.

```
Clear[frame];
frame[0] := Balloon;
                      (1 0 1)
frame[i_] := (frame[i] = (0 1 1) . frame[i - 1]);
                      (0 0 1)
Table[PlotImage[frame[i], 0, 100, 0, 60], {i, 0, 20}];
```

Exercise 2: Now it is your turn to create a motion picture. Execute a sequence of four commands similar to the commands used to create the frames of the last motion picture which assigns **frame[0]** to be the image of **Balloon** after it has been translated 40 units right and 30 units up. Then create a sequence of 21 frames showing the balloon moving one unit right and $\frac{3}{4}$ of a unit downward from one frame to the next.

After the frames have been created, animate your movie and then delete the picture frames.

Exercise 3: Reexecute the previous commands after making the following changes: Reassign **frame[0]** so that it represents a balloon whose bottom left hand corner is located at the point (10,30). Then create a sequence of 31 frames where each frame is located $\frac{1}{2}$ of a unit down and $\frac{3}{4}$ of a unit to the right of the previous frame. After the frames have been created, animate your movie and then delete the picture frames.

Defining Frames Piecewise

Suppose you want the balloon to rise up into the air, then move horizontally for a while and then come back down to the ground. In this case, the frames must be defined piecewise.

Exercise 4: Study the following two assignment statements:

$$\mathbf{frame[i\_] := \left(frame[\mathit{i}] = \begin{pmatrix} 1 & 0 & 1 \\ 0 & 1 & 0 \\ 0 & 0 & 1 \end{pmatrix} . frame[\mathit{i}-1]\right) /;\quad 16 \le \mathit{i} \le 25;}$$

and

$$\mathbf{frame[i\_] := \left(frame[\mathit{i}] = \begin{pmatrix} 1 & 0 & 1 \\ 0 & 1 & -1 \\ 0 & 0 & 1 \end{pmatrix} . frame[\mathit{i}-1]\right) /;\quad 26 \le \mathit{i} \le 40;}$$

Explain the purpose of each assignment statement here:

Once you have completed your explanation, check your work by executing the following input cell.

```
Clear[frame];
frame[0] := Balloon;
```

$$\text{frame[i\_]} := \left(\text{frame[i]} = \begin{pmatrix} 1 & 0 & 1 \\ 0 & 1 & 1 \\ 0 & 0 & 1 \end{pmatrix} . \text{frame[i - 1]}\right) \text{ /; } 1 \le i \le 15;$$

$$\text{frame[i\_]} := \left(\text{frame[i]} = \begin{pmatrix} 1 & 0 & 1 \\ 0 & 1 & 0 \\ 0 & 0 & 1 \end{pmatrix} . \text{frame[i - 1]}\right) \text{ /; } 16 \le i \le 25;$$

$$\text{frame[i\_]} := \left(\text{frame[i]} = \begin{pmatrix} 1 & 0 & 1 \\ 0 & 1 & -1 \\ 0 & 0 & 1 \end{pmatrix} . \text{frame[i - 1]}\right) \text{ /; } 26 \le i \le 40;$$

```
Table[PlotImage[frame[i], 0, 100, 0, 60], {i, 0, 40}];
```

Exercise 5: Create **frame[0]** through **frame[40]** so that:

a. **Balloon** is assigned to **frame[0];**

b. **frame[1]** through **frame[10]** are associated with frames of the balloon rising vertically 1 unit at a time;

c. **frame[11]** through **frame[20]** are associated with frames of the balloon moving right 1 unit and dropping 1 unit from one frame to the next;

d. **frame[21]** through **frame[30]** are associated with frames of the balloon moving right 1 unit and rising 1 unit from one frame to the next;

e. **frame[31]** through **frame[40]** are associated with frames of the balloon dropping one unit from one frame to the next.

After the frames have been created, animate your movie and then delete the picture frames.

Creating More Sophisticated Movies

Combining Rotations, Scaling and Translations

Combining translations, rotations and scalings can lead to more complex movies as will be seen in this section.

The image corresponding to **jetmat** is a jet similar to the image encountered in the previous assignment. Execute the following command to see the image.

```
PlotImage[jetmat, -20, 20, -20, 20];
```

As you discovered in the previous assignment, the matrix $\begin{pmatrix} \text{Cos}[\theta] & -\text{Sin}[\theta] & a - a\,\text{Cos}[\theta] + b\,\text{Sin}[\theta] \\ \text{Sin}[\theta] & \text{Cos}[\theta] & b - b\,\text{Cos}[\theta] - a\,\text{Sin}[\theta] \\ 0 & 0 & 1 \end{pmatrix}$ is used to rotate an image counterclockwise about the point (a, b) at an angle of θ. The *Mathematica* command **Rot[θ, a, b]** will be used to represent this matrix. For example, study and then execute the following input cell to create frames of a movie which rotates the jet about its tail $(-12, 0)$ four degrees ($\frac{\pi}{45}$ radians) at a time for 15 frames.

```
Clear[frame];
frame[0] = jetmat;
frame[i_] := (frame[i] = Rot[π/45, -12, 0] . frame[i - 1]);
Table[PlotImage[frame[i], -20, 20, -20, 20], {i, 0, 15}];
```

Exercise 6: Create a movie of the jet first rotating 5 degrees at a time counterclockwise about the point (-12,0) until it reaches a 45 degree angle and after the rotation, have the jet move 1 unit right and 1 unit up in frames 10 through 30. After the frames have been created, animate your movie and then delete the picture frames.

Scaling

In the next example, the goal is to create a movie using translations and scalings to give a balloon the appearance of rising into the air and moving away from the viewer. Before doing this, a brief review of scaling is needed.

The function **Scale[a,b,s,t]** represents a 3×3 matrix $\begin{pmatrix} s & 0 & a\,(1-s) \\ 0 & t & b\,(1-t) \\ 0 & 0 & 1 \end{pmatrix}$ used to scale an image about the point (a,b). For example, the product **Scale[0, 0, .5, .5] . Balloon** is associated with the image where the distance of each x-coordinate in **Balloon** from x=0 is cut in half and the distance of each y-coordinate from y=0 is cut in half as illustrated in the following figure where the image of the balloon appears on the left and the image of **Scale[0,0,.5, .5].Balloon** appears on the right.

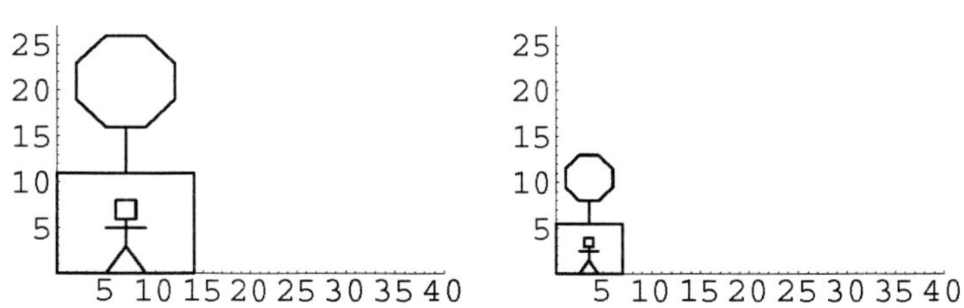

On the other hand, **Scale[14, 11, .5, .5] . Balloon** represents an image where the distance of each x-coordinate in Balloon from x=14 is cut in half while the distance of each y-coordinate from y=11 is cut in half. See the following figure.

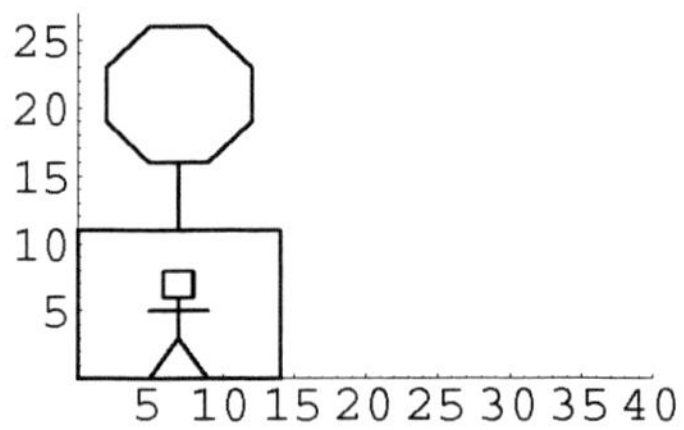

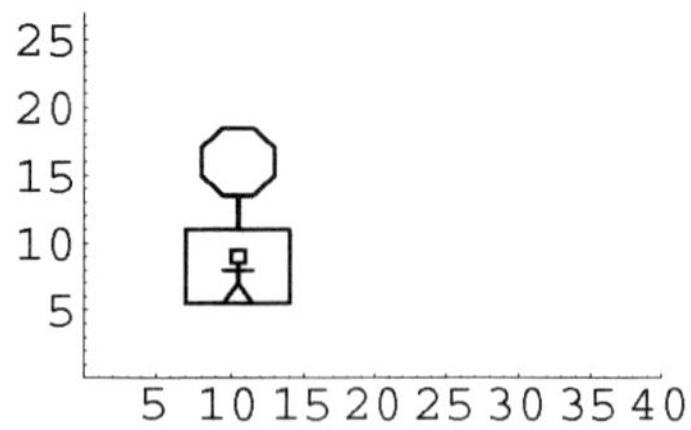

So for the matrix **Scale[a,b,s,t]** we call (a,b) the *center of scaling.*

Using Scaling and Translations

The goal of the example to follow is to create a movie of a balloon rising and drifting away from the viewer. The basic step, which is repeated many times, is to reduce the size of the balloon to 90% of its original size, using the bottom-left corner as the center of scaling, and then to move the balloon 1 unit right and 1 unit up. Initially, the corner is at (0, 0). The scaling has no effect on the corner, and the translation moves it to (1, 1). The next time the corner moves to (2, 2), and so on. The step to construct the movie are listed below.

1. The **Clear** command is applied to delete any previous defined values of frame.

2. Matrix **frame[0]** is assigned to equal **Balloon**.

3. The matrix $\begin{pmatrix} 1 & 0 & 1 \\ 0 & 1 & 1 \\ 0 & 0 & 1 \end{pmatrix}$ and **Scale[*i* – 1, *i* – 1, .9, .9]** are used to shrink the size of the balloon and to then move the balloon. The center of scaling used to create **frame[*i*]** is the bottom left hand corner of the image associated with **frame[*i* – 1]**, which in this case is the point (***i*** **– 1, *i*** **– 1**).

4. A **Table** command is used to plot the images of **frame[0]** through **frame[30]**.

Execute the following cell. After the frames have been created, animate your movie and then delete the picture frames.

```
Clear[frame];
frame[0] := Balloon;

frame[i_] := (frame[i] =

  (1 0 1)
  (0 1 1) . Scale[i - 1, i - 1, .9, .9] . frame[i - 1]);
  (0 0 1)
Table[PlotImage[frame[i], 0, 60, 0, 40], {i, 0, 30}];
```

The x-coordinate of the bottom left corner of **frame[*i* – 1]** is in the first row and first column of the matrix. This value can be obtained by executing **frame[*i* – 1][1, 1]]**. Similarly, the y-coordinate of the bottom left corner is in row 2, column 1 and is represented by **frame[*i* – 1][[2, 1]]**. So instead of using the command **Scale[i-1,1-1,.9, .9]** as in the previous input cell, it is better to use

Scale[frame[i-1][[1,1]],frame[i-1][[2,1]],.9, .9].

Execute the following input cell and compare it to the previous input cell.

```
Clear[frame];
frame[0] := Balloon;

frame[i_] := (frame[i] =

  (1 0 1)
  (0 1 1) .
  (0 0 1)

    Scale[frame[i - 1][[1, 1]], frame[i - 1][[2, 1]], .9, .9] . frame[i - 1]);

Table[PlotImage[frame[i], 0, 60, 0, 40], {i, 0, 30}];
```

Exercise 7: Create a movie where the balloon rises 1 unit vertically for 20 frames while it is reduced to 98% of the length and 98% of width of the previous frame. In the next 20 frames after that, have the balloon move right 1 unit at a time with the size of the balloon remaining constant. After the frames have been created, animate your movie and then delete the picture frames.

Exercise 8: Revise your movie in Exercise 8 so that as the balloon moves horizontally (after it stops rising), the balloon rotates counterclockwise 2 degrees about the bottom left hand corner of the image in each frame. After the frames have been created, animate your movie and then delete the picture frames.

Creating a Jet Movie

The purpose of this final example is to create a plane taking off and rising into the air and then finally leveling off after adequate altitude has been gained. To rotate a plane by an angle of θ about the bottom left corner of the plane and to then move the plane r units at an angle of θ requires the matrices $\begin{pmatrix} 1 & 0 & r\,\mathrm{Cos}[\theta] \\ 0 & 1 & r\,\mathrm{Sin}[\theta] \\ 0 & 0 & 1 \end{pmatrix}$ and **Rot[θ,a,b]** (see the figure below).

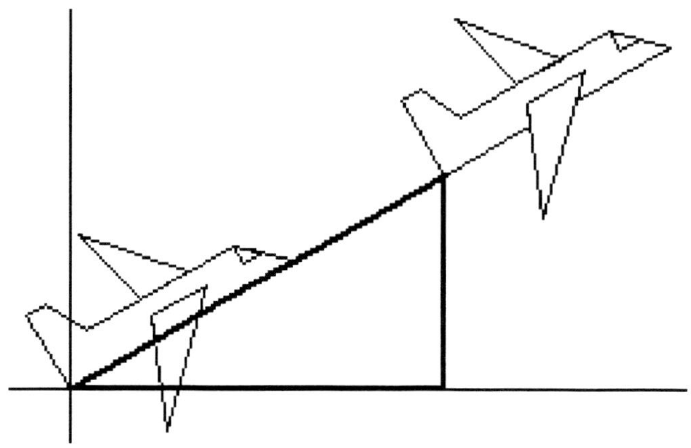

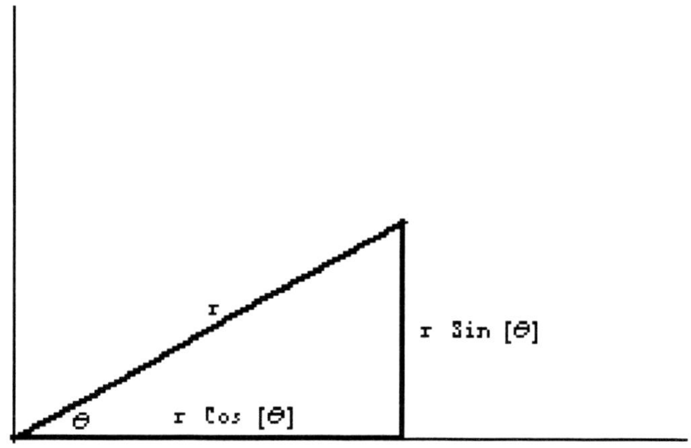

Execute the following input cell after first studying what each of the following five statements in the cell accomplish:

1. The **Clear** command is used.

2. Matrix **frame[0]** is set equal to **jetmat**.

3. For $i = 1, 2, \ldots, 10$, **frame[*i*]** is created by first rotating the image associated with **frame[*i* – 1]** about the bottom left-hand corner of its tail by an angle of 2 degrees counterclockwise ($\frac{\pi}{90}$ radians) and then translating the image 3 units at an angle of $\frac{i\pi}{90}$ radians. Note that the bottom left-hand corner of the tail of the image represented by **frame[*i* – 1]** is in the 10th column of **frame[*i* – 1].**

4. For $i = 11, 12, \ldots, 20$, **frame[*i*]** is created by first rotating the image associated with **frame[*i* – 1]** about the bottom left-hand corner of its tail by an angle of 2 degrees clockwise ($\frac{\pi}{90}$ radians) and then translating the image 3 units at an angle of $\frac{(20-i)\pi}{90}$ radians.

5. The frames of the movie are created.

After viewing the frames created by the following input cell, animate your movie and then delete the picture frames.

```
Clear[frame];
frame[0] := jetmat;

frame[i_] := (frame[i] = ({{1, 0, 3 Cos[iπ/90]}, {0, 1, 3 Sin[iπ/90]}, {0, 0, 1}}) .
    Rot[π/90, frame[i - 1][[1, 10]], frame[i - 1][[2, 10]]] . frame[i - 1]) /;
  1 ≤ i ≤ 10;

frame[i_] := (frame[i] = ({{1, 0, 3 Cos[(20-i)π/90]}, {0, 1, 3 Sin[(20-i)π/90]}, {0, 0, 1}}) .
    Rot[-π/90, frame[i - 1][[1, 10]], frame[i - 1][[2, 10]]] . frame[i - 1]) /;
  11 ≤ i ≤ 20;

frame[i_] := (frame[i] = ({{1, 0, 3}, {0, 1, 0}, {0, 0, 1}}) . frame[i - 1]) /; 21 ≤ i ≤ 30;
Table[PlotImage[frame[i], -15, 90, -10, 35], {i, 0, 30}];
```

Exercise 9: Copy and paste the lines of code just executed to a cell immediately below this exercise statement. Add new commands so the plane will eventually land back on the x axis. After the frames have been created, animate your movie and then delete the picture frames. You will need to adjust the horizontal range of values in the **PlotImage** command.

Initialization Files

The following commands will be executed automatically - do not reexecute the following commands.

```
Off[General::spell1];
```

```
$Post := If[ListQ[#], MatrixForm[#], #]&;
```

```
Needs["LinearAlgebra`MatrixManipulation`"];
```

```
HGMatrix[mat_] := Transpose[AppendRows[mat, Table[{1}, {Length[mat]}]]];
```

```
PlotImage[mat_, xmin_, xmax_, ymin_, ymax_] :=
 ListPlot[Transpose[TakeRows[mat, 2]], PlotJoined -> True, Axes -> Automatic,
  AxesOrigin -> {0, 0}, PlotRange -> {{xmin, xmax}, {ymin, ymax}}];
```

$$\text{Rot}[\theta\_] := \begin{pmatrix} \text{Cos}[\theta] & -\text{Sin}[\theta] & 0 \\ \text{Sin}[\theta] & \text{Cos}[\theta] & 0 \\ 0 & 0 & 1 \end{pmatrix};$$

$$\text{Rot}[\theta\_, a\_, b\_] := \begin{pmatrix} \text{Cos}[\theta] & -\text{Sin}[\theta] & a - a\,\text{Cos}[\theta] + b\,\text{Sin}[\theta] \\ \text{Sin}[\theta] & \text{Cos}[\theta] & b - b\,\text{Cos}[\theta] - a\,\text{Sin}[\theta] \\ 0 & 0 & 1 \end{pmatrix};$$

$$\text{Scale}[s\_, t\_] := \begin{pmatrix} s & 0 & 0 \\ 0 & t & 0 \\ 0 & 0 & 1 \end{pmatrix};$$

$$\text{Scale}[a\_, b\_, s\_, t\_] := \begin{pmatrix} s & 0 & a\,(1-s) \\ 0 & t & b\,(1-t) \\ 0 & 0 & 1 \end{pmatrix};$$

$$\text{YRef} = \begin{pmatrix} -1 & 0 & 0 \\ 0 & 1 & 0 \\ 0 & 0 & 1 \end{pmatrix};$$

```
Balloon = Transpose[{{0, 0, 1}, {5, 0, 1}, {7, 3, 1}, {7, 5, 1}, {5, 5, 1}, {9, 5, 1},
    {7, 5, 1}, {7, 6, 1}, {6, 6, 1}, {6, 8, 1}, {8, 8, 1}, {8, 6, 1}, {7, 6, 1},
    {7, 3, 1}, {9, 0, 1}, {4, 0, 1}, {14, 0, 1}, {14, 11, 1}, {0, 11, 1}, {0, 0, 1},
    {0, 11, 1}, {7, 11, 1}, {7, 16, 1}, {9, 16, 1}, {12, 19, 1}, {12, 23, 1},
    {9, 26, 1}, {5, 26, 1}, {2, 23, 1}, {2, 19, 1}, {5, 16, 1}, {7, 16, 1}}];
```

```
jet = {{0, 0}, {5, 0}, {9, -4}, {7, 1}, {4, 1}, {5, 0}, {4, 1}, {7, 1},
   {7.4, 0}, {12, 0}, {12, 4}, {11, 4}, {10, 2}, {7, 2}, {8, 6}, {4.5, 2 },
   {7, 2}, {2, 2}, {2, 1}, {1, 1}, {2, 2}, {0, 0}};
```

```
jetmat = YRef . HGMatrix[jet];
```

Computer Assignment: Properties of Determinants

Purpose: To examine properties of determinants and to study the determinants of special matrices.

***Mathematica* Functions:** Clear, Det, Factor, IdentityMatrix, Inverse, RandMat, ReplaceRow, Scale, Swap

Palette Used in this Assignment:

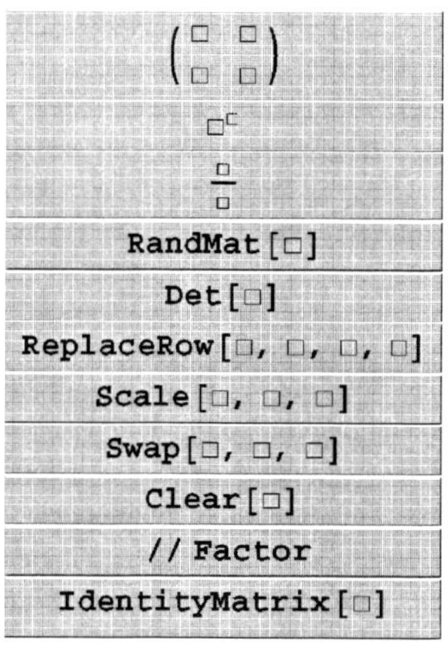

Introduction: As will be seen in Chapter 3, knowing some of the properties of determinants can greatly simplify determinant computations. This assignment will allow you to discover some of these properties with the help of *Mathematica*.

Row Operations and Determinants

The command **RandMat[n]** creates an n×n matrix whose entries are random integers between −9 and 9. Execute the following assignment statement.

```
A = RandMat[5]
```

Execute the following command to compute the determinant of *A*.

```
Det[A]
```

Interchanging Rows

Suppose you want to find the determinant of the matrix obtained by switching rows 1 and 2 of *A*. Execute the following command.

```
Swap[A, 1, 2]
```

Keep in mind that matrix *A* has not changed! **Swap** (as well as the **ReplaceRow** and **Scale**) actually creates a new matrix and leaves *A* unchanged. Now compute the determinant of this new matrix by executing the following command after first entering the associated output number.

```
Det[% □]
```

Exercise 1: For each part below, compute the determinant of the new matrix and compare its determinant to *A*.

a. Swap rows 1 and 3 of *A*.

b. Swap rows 2 and 5 of *A*.

c. Swap rows 4 and 5 of *A*.

Exercise 2: Based upon the results of Exercise 1, complete the following conjecture by filling in the box:

If two rows of an n×n matrix A are interchanged to produce matrix B, then Det[B]= □ Det[A]

Using the ReplaceRow Command

Execute the following two commands to first create a new random 4×4 matrix and to then compute its determinant.

```
A = RandMat[4]
```

```
Det[A]
```

Recall that the command **ReplaceRow[A,r,m,s]**replaces row r of matrix A by its sum with m times row s. Execute the following command.

```
ReplaceRow[A, 1, 5, 3]
```

Now compute the determinant of this matrix using the following input cell.

```
Det[% □]
```

Exercise 3: For each part below, compute the new matrix and then compare its determinant with the determinant of *A*.

a. ReplaceRow[A,2,7,3]

b. ReplaceRow[A,4,-6,2]

c. ReplaceRow[A,3,-4,2]

Exercise 4: Based upon the results of Exercise 3, fill in the missing blank in the following conjecture:

If a multiple of one row of an n×n matrix A is added to another row of A to produce a matrix B, then Det[B]=□

To verify your conjecture in Exercise 4 for the 3×3 case, begin by executing the following commands. (The semicolon separates the two commands and it also suppresses any output from being displayed.)

$$\text{Clear[a, b, c, d, e, f, g, h, i]; A} = \begin{pmatrix} a & b & c \\ d & e & f \\ g & h & i \end{pmatrix};$$

To find a formula for the determinant of a 3×3 matrix, compute the following.

```
Det[A]
```

Execute the following command so that row 1 of *A* is replaced by its sum with *m* times row 2.

```
ReplaceRow[A, 1, m, 2]
```

Use the following input cell to find the determinant of this new matrix.

```
Det[% □]
```

To compare the value of this determinant to det *A*, subtract the two values. In the following input cell, enter the output number associated with det *A* in the first box and then enter the output number associated with the determinant just computed in the other box.

```
% □ - % □
```

Exercise 5: For each part below, compute the new matrix and then compare its determinant with the determinant of *A*.

a. ReplaceRow[A,1,m,3] b. ReplaceRow[A,2,m,3]

Using the Scale Command

Recall that the command **Scale[A, r, c]** multiplies row *r* of matrix *A* by a nonzero scalar *c*. Continuing on with the 3×3 matrix *A* used in Exercise 5, compute the following **Scale** command.

```
Scale[A, 1, 5]
```

The command **//Factor** will factor out any common terms contained in the determinant. Execute the following command to compute the determinant of the matrix obtained in the last output.

```
Det[% □] // Factor
```

Exercise 6: For each part below, compute the new matrix and then compare its determinant with the determinant of A.

a. Scale[A,2,-3] b. Scale[A,3,4]

Exercise 7: Based upon your work to Exercise 6, complete the following conjecture. (Fill in the blank box.)

If one row an n×n matrix A is multiplied by a constant r to obtain matrix B, then

Det[B]=□ Det[A].

Exercise 8: a. Based upon your answer to Exercise 7, complete the following conjecture.

If every row an an n×n matrix A is multiplied by a constant r to obtain matrix B, then

Det[B]=□ Det[A]. Explain your answer.

b. Execute the following cell to first clear our any previously defined values of *r* and to then compute the determinant of *r A* to verify that the conjecture stated in part a holds for the 3×3 case.

```
Clear[r]; Det[r A]
```

The Determinant of a Product

Execute the following two commands to create two random 4×4 matrices.

```
A = RandMat[4]
```

```
B = RandMat[4]
```

Execute the following command to compute the determinant of the product of *A* and *B*. (Recall that a matrix product *AB* is computed with *Mathematica* by inserting a period between *A* and *B*.)

```
Det[A . B]
```

Now compute the product of the determinants. (Since **Det[A]** and **Det[B]** are scalars, a space is place between them to indicate scalar multiplication.)

```
Det[A] Det[B]
```

Exercise 9: Form a conjecture describing the relationship between the determinants of *A*, *B* and *AB* where *A* and *B* are n×n.

Type in your conjecture here:

The Determinant of the Identity Matrix and the Inverse

The *Mathematica* command used to create an n×n identity matrix is **IdentityMatrix[n].** For example, execute the following command to create the 7×7 identity matrix.

```
IdentityMatrix[7]
```

Now compute its determinant.

```
Det[% □]
```

Exercise 10: Compute the determinants of the 3×3, 6×6 and 10×10 identity matrices. Make a conjecture about the determinant of any n×n identity matrix.

Exercise 11: Since $AA^{-1} = I$, then state the value of det AA^{-1}.

Type in your answer here and justify your answer:

Exercise 12: Since det AA^{-1} = (det A) (det A^{-1}), describe the relationship between det A and det A^{-1}.

Type in your answer here (with justification):

Exercise 13: Use **RandMat** to create a 6×6 random matrix A . Verify your answer to Exercise 12 by computing the determinants of A and A^{-1}.

Exercise 14: Explain why an invertible matrix cannot have a determinant equal to zero.

Type in your explanation here:

Initialization Cell

```
Needs["LayData`LayFunctions`"];
$Post := If[MatrixQ[#] || VectorQ[#], MatrixForm[#], #]&;
RandMat[n_] := Table[Random[Integer, {-9, 9}], {n}, {n}]
```

Computer Assignment: Markov Chains

Purpose: To investigate steady-state vectors associated with Markov chains.

Prerequisite Reading: Section 4.9; read about the **NulBasis** command on page MM-14 in the *Study Guide*.

***Mathematica* Functions:** Clear, MatrixPower, N, NulBasis, ProbVec, RandStoch, $\sum_{\square=\square}^{\square}\square$, Table, IdentityMatrix

Palette Used in this Assignment:

`Table[□, {□, □, □}]`
`Clear[□]`
`x[0] = {□}`
`x[k_] :=`
`x[□]`
$\begin{pmatrix}\square & \square \\ \square & \square\end{pmatrix}$
$\begin{pmatrix}\square \\ \square\end{pmatrix}$
`IdentityMatrix[□]`
`NulBasis[□]`
`N[□]`
`MatrixPower[□, □]`
`RandStoch[□]`
`ProbVec[□]`
$\sum_{\square=\square}^{\square}\square$
$\frac{\square}{\square}$

Introduction: The purpose of this assignment is to reinforce the concepts presented in Section 4.9. Important concepts to be familiar with before beginning this assignment include a *probability vector, stochastic matrix, Markov chain,* and a *steady-state vector.*

Forming a Markov Chain

Consider the opening example in Section 4.9 which examines the population of a particular city and the surrounding suburbs. Note that $\mathbf{x}_0 = \begin{pmatrix} .600 \\ .400 \end{pmatrix}$ where .600 represents the fraction of the population living in the city and .400 represents the fraction of the population living in the surrounding suburbs of the city in 1990.

To determine the population distribution for 1991, the *migration matrix* $P = \begin{pmatrix} .95 & .03 \\ .05 & .97 \end{pmatrix}$ is used to compute

$$\mathbf{x}_1 = \mathrm{P}\mathbf{x}_0 = \begin{pmatrix} .95 & .03 \\ .05 & .97 \end{pmatrix} \begin{pmatrix} .600 \\ .400 \end{pmatrix} = \begin{pmatrix} .582 \\ .418 \end{pmatrix}.$$

So in 1991, about 58.2% of the population is in the city while 41.8% of the population is in the suburbs.

To compute the population distributions for 1992, 1993, 1994, ... , compute $\mathbf{x}_k = P\mathbf{x}_{k-1}$ for $i = 2$, 3, and so forth. You can quickly obtain these values with *Mathematica* by defining a function **x[*k*]** equaling $\mathbf{x}_k$ for $k = 0, 1, 2, \ldots$.

Four-Step Outline for Defining a Function *x*[*k*]

The following input cell will complete all of the following steps.

1. Clear out any previously defined values of **x[*k*]**.

2. Define the stochastic matrix *P*.

3. Assign **x[0]** to be the value of $\mathbf{x}_0$. The vector $\begin{pmatrix} .6 \\ .4 \end{pmatrix}$ is represented by {.6, .4} with *Mathematica.*

4. Define **x[*k*]** to be *P* times **x[*k* – 1]** (:= is a *delayed equals sign*, which is used in define functions recursively and the underscore must be placed immediately after the *k* on the left hand side of the equation).

Execute the following:

```
Clear[x];
P = ( .95  .03 );
    ( .05  .97 )
x[0] = {.6, .4};
x[k_] := P . x[k - 1]
```

Now you can find the value of **x[*k*]** for a given integer *k*. For example, execute the following two cells to see the values of x_5 and x_8.

```
x[5]
```

```
x[8]
```

Exercise 1: Find the value of $\boldsymbol{x}_{10}$.

Now execute the following to instruct *Mathematica* to display **x[1]**, **x[2]**, ..., **x[100]**. The output will be a 100×2 table whose kth row is **x[*k*]**.

```
Table[x[k], {k, 1, 100}]
```

Now suppose in 1990 that 85% of the population lives within the city limits and the remaining 15% of the population lives in the suburbs (that is, suppose $\boldsymbol{x}_0 = \begin{pmatrix} .85 \\ .15 \end{pmatrix}$). Execute the following command to see how this change affects $\boldsymbol{x}_1, \boldsymbol{x}_2, \ldots, \boldsymbol{x}_{100}$.

```
x[0] = {.85, .15};
Table[x[k], {k, 1, 100}]
```

Exercise 2: Now suppose the 1990 population consists of 75% of the people living in the city and 25% living in the suburbs. Display the values of $\boldsymbol{x}_1, \boldsymbol{x}_2, \ldots, \boldsymbol{x}_{100}$.

Exercise 3: Three different tables of values of $\boldsymbol{x}_1, \boldsymbol{x}_2, \ldots, \boldsymbol{x}_{100}$ have now been created. Review Theorem 18 on page 288 in your textbook and then compare the three tables to make an educated guess at the approximate value of the steady-state vector $\boldsymbol{q}$. (Round the values in your answer to three decimal places.)

Fill in the values here: $\mathbf{q}=\begin{pmatrix} \square \\ \square \end{pmatrix}$

Tracking Rental Car Distributions

Read Exercise 16 found in Section 4.9 where the given stochastic matrix is $\begin{pmatrix} .90 & .01 & .09 \\ .01 & .90 & .01 \\ .09 & .09 & .90 \end{pmatrix}$. At the beginning of the year, suppose Hertz distributes 50% of its Detroit fleet of cars at the city airport, 30% downtown and the remaining 20% at the metro airport. If you let $\boldsymbol{x}_0 = \begin{pmatrix} .5 \\ .3 \\ .2 \end{pmatrix}$ and $\boldsymbol{x}_k = \begin{pmatrix} .90 & .01 & .09 \\ .01 & .90 & .01 \\ .09 & .09 & .90 \end{pmatrix} \boldsymbol{x}_{k-1}$, then $\boldsymbol{x}_k$ represents the distribution of the cars on the kth day of the year.

Exercise 4: Use the four-step outline described earlier to define **x[*k*]** and then display the values $\mathbf{x}_1, \mathbf{x}_2, \ldots, \mathbf{x}_{100}$. Let P equal $\begin{pmatrix} .90 & .01 & .09 \\ .01 & .90 & .01 \\ .09 & .09 & .90 \end{pmatrix}$.

Exercise 5: Display $\mathbf{x}_1, \mathbf{x}_2, \ldots, \mathbf{x}_{100}$ for each initial distribution below and then make a conjecture about the value of $\mathbf{q}$ (rounding the entries to two decimal places and keep in mind that the values in $\mathbf{q}$ should add up to one):

(a) $\mathbf{x}_0 = \begin{pmatrix} .6 \\ .2 \\ .2 \end{pmatrix}$ (b) $\mathbf{x}_0 = \begin{pmatrix} .4 \\ .25 \\ 35 \end{pmatrix}$

Conjecture (enter the missing values) : $\mathbf{q} = \begin{pmatrix} \square \\ \square \\ \square \end{pmatrix}$

Exercise 6: The command **MatrixPower[P,n]** will computer P^n. Compute the value of P^5. Then compute P^n for $n = 10, 30, 50, 100$ and 200. What happens to the columns of P^n as n increases? For large n, how do the columns of P^n appear to relate to the value **q** in the previous exercise?

Enter your answer here:

Exercise 7: What eventually happens to the distribution of rental cars at the three locations and how does the value of $\mathbf{x}_0$ affect the eventual distribution?

Enter your answer here:

Determining the Exact Value of the Steady-State Vector

Example 5, beginning on page 287, outlines the steps for finding the steady-state vector $\mathbf{q}$ associated with a stochastic matrix P. You can use *Mathematica* to quickly find the value.

First, P is redefined using fractions. (When numbers are entered in fractional form instead of decimal form, *Mathematica* will perform exact arithmetic. If you do not need a final fractional answer, you can instead define P as in Exercise 4.) In the second line of the following input cell, the command **P – IdentityMatrix[3]** creates the matrix $P - I$, where I is the 3×3 identity matrix and **NulBasis** finds a solution to $(P - I)\mathbf{x} = \mathbf{0}$. The output of **NulBasis** matches what you would obtain if you made the calculations by hand.

```
     ⎛ 9/10    1/100   9/100 ⎞
P =  ⎜ 1/100   9/10    1/100 ⎟ ;
     ⎝ 9/100   9/100   9/10  ⎠
w = NulBasis[P - IdentityMatrix[3]]
```

Notice that $\boldsymbol{w} = \begin{pmatrix} \frac{91}{99} \\ \frac{19}{99} \\ 1 \end{pmatrix}$, is displayed in row form. Now you divide $\boldsymbol{w}$ by the sum of its entries to obtain $\boldsymbol{q}$. *Mathematica* notation for the *jth* entry in $\boldsymbol{w}$ is ***w*[[1, *j*]]**. The command $\sum_{j=1}^{3}$ ***w*[[1, *j*]]** will add up these entries. Execute the following command (a space between the fraction and ***w*** was inserted to indicate multiplication).

```
q = 1/(Σ_{j=1}^{3} w[[1, j]]) w
```

If you are interested in an approximate value of the steady-state vector, execute the command **N[q]** below.

```
N[q]
```

Additional Exercises

Exercise 8: Execute the following two commands to create a random 3×3 stochastic matrix P and to then create M, the decimal form of P. Compute M^n for $n = 5$, 10, and 20. Continue to compute M^n for larger and larger values of n, if necessary, until the entries stabilize to 3 decimal places. Then estimate the value of the steady-state vector **q**. *(Note: You may not be able to estimate **q** in the rare event where **RandStoch[3]** does not generate a regular stochastic matrix. In this case, start over by execute the following two statements again.)*

```
P = RandStoch[3]
```

```
M = N[P]
```

Exercise 9: The command **ProbVec[n]** creates a $n \times 1$ probability vector. Use this command to assign ***x*[0]** to be a random 3×1 probability vector. Compute the value of ***x*[*n*]** for $n = 1, 5$ and 10, using the value of M from Exercise 8. Continue to compute **x[n]** for larger and larger values of n until the entries in *x*[*n*] stabilize to three decimal places. Compare your results with the estimated value of ***q*** in the previous exercise.

Exercise 10: Use matrix *P* from Exercise 8 to find the exact value of ***q***. Use ***N*[*q*]** to compare your answer with the previous two exercises.

Initialization Cells

```
$Post := If[MatrixQ[#] || VectorQ[#], MatrixForm[#], #]&;

Needs["LayData`LayFunctions`"];

RandStoch[n_] := Module[{mat, sum},
  mat = Table[Random[Integer, {0, 100}]/100, {i, n}, {j, n}];
  sum = Fold[Plus, 0, mat];
  Return[Transpose[Transpose[mat] / sum]]];

ProbVec[n_] := Module[{vec, sum},
  vec = Table[Random[Integer, {0, 100}], {n}];
  sum = Fold[Plus, 0, vec];
  Return[vec / sum]]
```

Computer Assignment: Survival of the Spotted Owls

Purpose: To use eigenvalues to study population changes in the spotted owl example introduced in Chapter 5.

Prerequisite Reading: Introduction to Chapter 5, Section 5.1 and Section 5.6.

***Mathematica* Functions:** Abs, Clear, Eigenvalues, ListPlot, Show, Table.

Palette used in this assignment:

```
λ
λ = Eigenvalues[□]
Abs[□]
(□ □
 □ □)
x[k_] :=
x[□] = {□, □, □}
Clear[□]
Show[□]
Table[x[k][[□]], {k, 1, □}]
ListPlot[□,
  PlotRange → {□, □},
  PlotStyle → {PointSize[.02],
  RGBColor[□, □, □]}]
```

Introduction: To determine whether or not the spotted owl population is doomed to extinction, you will examine the eigenvalues of the associated stage matrix $\begin{pmatrix} 0 & 0 & .33 \\ t & 0 & 0 \\ 0 & .71 & .94 \end{pmatrix}$ for various values of t (where t represents the percentage of juvenile spotted owls that survive to become subadults). You will also determine the minimum value of t needed so that the spotted owl population avoids extinction. By working through this assignment, you will also learn how to quickly obtain eigenvalues of a given matrix and to plot the spotted owl populations over time.

Predicting the Fate of the Spotted Owl Population

The introduction to Chapter 5 in your textbook discusses the plight of the spotted owls in the Pacific Northwest. The population of spotted owls is divided into *juveniles, subadults* and *adults*. Since it is assumed that there is approximately the same number of males as females, only the females are counted in the study. Recall that j_k, s_k and a_k represent the number of female spotted owls in the juvenile, subadult and adult life stages, respectively, at year k. Using the stage matrix $A=\begin{pmatrix} 0 & 0 & .33 \\ .18 & 0 & 0 \\ 0 & .71 & .94 \end{pmatrix}$, the female population distribution, $(j_{k+1}, s_{k+1}, a_{k+1})$, at time k+1 can be determined by computing $\begin{pmatrix} j_{k+1} \\ s_{k+1} \\ a_{k+1} \end{pmatrix} = \begin{pmatrix} 0 & 0 & .33 \\ .18 & 0 & 0 \\ 0 & .71 & .94 \end{pmatrix} \begin{pmatrix} j_k \\ s_k \\ a_k \end{pmatrix}$.

Exercise 1: Complete parts a though d.

a. Express $\begin{pmatrix} j_{k+1} \\ s_{k+1} \\ a_{k+1} \end{pmatrix} = \begin{pmatrix} 0 & 0 & .33 \\ .18 & 0 & 0 \\ 0 & .71 & .94 \end{pmatrix} \begin{pmatrix} j_k \\ s_k \\ a_k \end{pmatrix}$ as a system of three equations and three unknowns by filling in the following missing values.

$j_{k+1} = \square\, j_k + \square\, s_k + \square\, a_k$

$s_{k+1} = \square\, j_k + \square\, s_k + \square\, a_k$

$a_{k+1} = \square\, j_k + \square\, s_k + \square\, a_k$

b. The top equation in part a is $j_{k+1} = .33\, a_k$ which implies that the number of new female juveniles in year k+1 is .33 times the number of adult females in year k. (As stated in the introduction to Chapter 5 on the top of page 296 in your textbook.) Explain the meaning of the second and third equations in part a and check your work with the information given in the textbook introduction to Chapter 5.

Write your explanation here:

c. From your reading of the introduction to Chapter 5, explain how the value of .18 was determined in the stage matrix.

Write your explanation here:

d. If instead, 70% of the juveniles were to survive to leave the nest and 40% of the juveniles that left the nest were able to find new home ranges, then what number would replace .18 in the stage matrix?

Enter the number here: □

Letting $\mathbf{x}_k = \begin{pmatrix} j_k \\ s_k \\ a_k \end{pmatrix}$, then $\mathbf{x}_{k+1} = A\mathbf{x}_k$. In Section 5.6, it is shown that $\mathbf{x}_k = c_1(\lambda_1)^k\,\mathbf{v}_1 + c_2(\lambda_2)^k\,\mathbf{v}_2 + c_3(\lambda_3)^k\,\mathbf{v}_3$ where λ_1, λ_2, and λ_3 are the eigenvalues of A with corresponding eigenvectors $\mathbf{v}_1$, $\mathbf{v}_2$ and $\mathbf{v}_3$, respectively. The command **Eigenvalues[A]**returns a list of eigenvalues for a given n×n matrix A. Execute the following cell to create A and to let λ represent the list of eigenvalues of A.

```
A = ( 0    0    .33 )
    ( .18  0    0   ) ;
    ( 0    .71  .94 )
λ = Eigenvalues[A]
```

Execute the following command to see the absolute value of each eigenvalue, respectively.

```
Abs[λ]
```

Note that all the eigenvalues in this example have absolute values less than one. Therefore $c_1(\lambda_1)^k\,\mathbf{v}_1$, $c_2(\lambda_2)^k\,\mathbf{v}_2$, and $c_3(\lambda_3)^k\,\mathbf{v}_3$ all tend to zero as $k \to \infty$ and since

$$\mathbf{x}_k = c_1(\lambda_1)^k\,\mathbf{v}_1 + c_2(\lambda_2)^k\,\mathbf{v}_2 + c_3(\lambda_3)^k\,\mathbf{v}_3, \tag{1}$$

$\mathbf{x}_k$ tends to $\begin{pmatrix} 0 \\ 0 \\ 0 \end{pmatrix}$ as $k \to \infty$.

Now suppose the value of .18 in the stage matrix is replaced by t so that $A = \begin{pmatrix} 0 & 0 & .33 \\ t & 0 & 0 \\ 0 & .71 & .94 \end{pmatrix}$. Execute the following cell to clear out any value that may have been previously assigned to t and to redefine the stage matrix A.

```
Clear[t];
A = ( 0  0    .33 )
    ( t  0    0   ) ;
    ( 0  .71  .94 )
```

You have already seen that the owl population declines towards extinction when $t = .18$. To consider the case where $t = .32$, execute the following assignment statement to assign this new value to t and to see the associated stage matrix.

```
t = .32; A
```

Exercise 2: Determine the eigenvalues of the new stage matrix just created. Then find the absolute values of the eigenvalues using the **Abs** command.

Exercise 3: Use the information obtained in Exercise 2 to complete this exercise.

a. The eigenvalue with the largest absolute value is called the **dominant** eigenvalue. Enter the dominant eigenvalue in the box below.

Dominant eigenvalue: □

b. Suppose λ_1 represents the dominant eigenvalue of the stage matrix in the vector decomposition (1) above. Explain why $\mathbf{x}_k$ does not tend to zero if $c_1 \neq 0$.

Enter your explanation here:

c. For large k, explain why $\mathbf{x}_k \approx c_1(\lambda_1)^k\,\mathbf{v}_1$, where λ_1 is the dominant eigenvalue.

Enter your explanation here:

d. Explain why the owl population will grow slowly and give the long-term growth rate of the population.

Enter your explanation here:

Exercise 4: Complete each part below.

a. Change the value of t to .19 and then determine the absolute values of the eigenvalues of A. Will the owl population still decline in this case? (Explain your reasoning below.)

Explanation:

b. Reexecute the input cells you created to complete part a after first changing the value of t to .20. Will the owl population still decline? (Explain.)

Explanation:

c. Find the dominant eigenvalue for $t = .21$, .22, .23 and so forth until a dominant eigenvalue with an absolute value larger than 1 is obtained. Call this value of $t = t_0$ and fill in this value below. (Hint: Continue to reexecute the input cells used in parts a and b after first changing the value of t in each case.)

Answer (fill in): □

d. For large k, recall that $\mathbf{x}_k \approx c_1(\lambda_1)^k\,\mathbf{v}_1$, where λ_1 is the dominant eigenvalue. If $c_1 \neq 0$, then explain the significance of finding a dominant eigenvalue larger than 1.

Explanation:

Plotting the Owl Populations

You can graph year-by-year populations of spotted owls after first executing the following cell which contains four assignment statements. Each of these assignment statements is explained here:

1. **Clear[x]** clears out any previously defined values of function ***x***.

2. The parameter *t* is reset to equal to 0.18.

3. Set ***x*[0]** equal to $\mathbf{x}_0$, the initial number of owls in each age group. (We will assume that there are initially 1000 owls in each age group for the sake of simplicity.)

4. Let ***x*[*k*]** represent x_k and set the function ***x*[*k*]** equal to *A* times ***x*[*k* – 1]**. This is a recursive equation and a delayed equals (:=) must therefore be used. The underscore after the *k* on the left hand side of the assignment statement lets *Mathematica* know that *k* is a variable.

```
Clear[x];
t = .18;
x[0] = {1000, 1000, 1000};
x[k_] := A.x[k - 1];
```

The value of ***x*[*k*]** will be a vector and ***x*[*k*] [[*i*]]** represents the ith entry in the vector. Therefore ***x*[*k*] [[*1*]]** represents the juvenile population in year *k*. The following assignment statement uses the **Table** command to list the juvenile population for years 1 through 50

```
juvpop = Table[x[k][[1]], {k, 1, 50}]
```

Finally, you can plot the juvenile population by executing the following assignment statement. The command **PlotRange->{0,1700}** is used to create a vertical axis from 0 to 1700 and

PlotStyle->{PointSize[.02],RGBColor[1,0,0]}

is used to control the size of the points being plotted and to plot the points in the color red (**RGBColor[1,0,0]** represents red, **RGBColor[0,1,0]** represents green, and **RGBColor[0,0,1]** represents blue). Execute the following command.

```
juvplot = ListPlot[juvpop, PlotRange -> {0, 1700},
  PlotStyle -> {PointSize[.02], RGBColor[1, 0, 0]}]
```

⁂Exercise 5: Use the **Table** command to create a list of the subadult populations for years 1 through 50 and assign this list the name **subpop**. Create a plot of the subadult populations, called **subplot**, which plots the list data in green. Use the same **PlotRange** values as given in **juvplot**.

⁂Exercise 6: Use the **Table** command to create a list of the adult populations for years 1 through 50 and assign this list the name **adultpop**. Create a plot of the adult populations, called **adultplot**, which plots the list data in blue. Use the same **PlotRange** values as given in **juvplot**.

⁂Exercise 7: If you have successfully completed plots called juvplot, subplot and adultplot, then execute the following command to see all three plots simultaneously. Do these graphs agree with your expectations based upon your earlier work? (Explain your reasoning.)

Explain here:

```
Show[juvplot, subplot, adultplot]
```

⁂Exercise 8: Create plots for the juvenile, subadult and adult populations when $t = .32$. Also show all three plots simultaneously. Do these graphs agree with your expectations based upon your earlier work? (Explain your reasoning.) *(Hint: Instead of entering a series of new input cells, just set t=.30 immediately after this cell and then scroll up the page and reexcute the previous commands, starting with the assignment statement for juvpop to obtain the new graphs.*

⁂Exercise 9: Create plots for the juvenile, subadult and adult populations for the value of $t = t_0$ found in Exercise 4c. Also show all three plots simultaneously. The hint in Exercise 8 applies to this exercise also!

What do the graphs indicate is happening to the populations in this case?

Explain here:

Initialization Cell

```
$Post := If[MatrixQ[#], MatrixForm[#], #]&
```